# ACT Aspire Subject Test

## Mathematics Grade 8

### Student Practice Workbook

### + Two Full-Length ACT Aspire Math Tests

Math Notion

www.MathNotion.com

# ACT Aspire Subject Test Mathematics Grade 8

ACT Aspire Subject Test Mathematics Grade 8

Published in the United State of America By

The Math Notion

Web: WWW.MathNotion.com

Email: info@Mathnotion.com

Copyright © 2021 by the Math Notion. All rights reserved. No part of this publication may be reproduced, stored in a retrieval system, or transmitted in any form or by any means, electronic, mechanical, photocopying, recording, scanning, or otherwise, except as permitted under Section 107 or 108 of the 1976 United States Copyright Ac, without permission of the author.

All inquiries should be addressed to the Math Notion.

ISBN: 978-1-63620-069-9

**ACT Aspire Subject Test Mathematics Grade 8**

# The Math Notion

**Michael Smith** has been a math instructor for over a decade now. He launched the Math Notion. Since 2006, we have devoted our time to both teaching and developing exceptional math learning materials. As a test prep company, we have worked with thousands of students. We have used the feedback of our students to develop a unique study program that can be used by students to drastically improve their math scores fast and effectively. We have more than a thousand Math learning books including:

- **SAT Math Prep**
- **ACT Math Prep**
- **SSAT/ISEE Math Prep**
- **Accuplacer Math Prep**
- **Common Core Math Prep**
- **many Math Education Workbooks, Study Guides, Practice and Exercise Books**

As an experienced Math test preparation company, we have helped many students raise their standardized test scores—and attend the colleges of their dreams: We tutor online and in person, we teach students in large groups, and we provide training materials and textbooks through our website and through Amazon.

You can contact us via email at:

info@Mathnotion.com

# ACT Aspire Subject Test Mathematics Grade 8

**Get the Targeted Practice You Need to Ace the ACT Aspire Math Test!**

**ACT Aspire Subject Test Mathematics Grade 8** includes easy-to-follow instructions, helpful examples, and plenty of math practice problems to assist students to master each concept, brush up their problem-solving skills, and create confidence.

The ACT Aspire math practice book provides numerous opportunities to evaluate basic skills along with abundant remediation and intervention activities. It is a skill that permits you to quickly master intricate information and produce better leads in less time.

Students can boost their test-taking skills by taking the book's two practice ACT Aspire Math exams. All test questions answered and explained in detail.

**Important Features of the 8th grade ACT Aspire Math Book:**

- **A complete review** of ACT Aspire math test topics,
- Over 2,500 practice problems covering all topics tested,
- The most important concepts you need to know,
- Clear and concise, easy-to-follow sections,
- Well designed for enhanced learning and interest,
- Hands-on experience with all question types,
- **2 full-length practice tests** with detailed answer explanations,
- Cost-Effective Pricing,

Powerful math exercises to help you avoid traps and pacing yourself to beat the ACT Aspire test. Students will gain valuable experience and raise their confidence by taking math practice tests, learning about test structure, and gaining a deeper understanding of what is tested on the ACT Aspire Math. If ever there was a book to respond to the pressure to increase students' test scores, this is it.

# ACT Aspire Subject Test Mathematics Grade 8

# WWW.MathNotion.COM

… So Much More Online!

- ✓ FREE Math Lessons

- ✓ More Math Learning Books!

- ✓ Mathematics Worksheets

- ✓ Online Math Tutors

For a PDF Version of This Book

Please Visit WWW.MathNotion.com

# ACT Aspire Subject Test Mathematics Grade 8

# Contents

**Chapter 1 : Integers and Number Theory ..........................................11**
   Rounding ..........................................................................................12
   Rounding and Estimates ..................................................................13
   Adding and Subtracting Integers .....................................................14
   Multiplying and Dividing Integers ...................................................15
   Order of Operations ........................................................................16
   Ordering Integers and Numbers ......................................................17
   Integers and Absolute Value ...........................................................18
   Factoring Numbers ..........................................................................19
   Greatest Common Factor .................................................................20
   Least Common Multiple ..................................................................21
   Answers of Worksheets ...................................................................22

**Chapter 2 : Fractions and Decimals ..................................................25**
   Simplifying Fractions ......................................................................26
   Adding and Subtracting Fractions ...................................................27
   Multiplying and Dividing Fractions .................................................28
   Adding and Subtracting Mixed Numbers ........................................29
   Multiplying and Dividing Mixed Numbers .....................................30
   Adding and Subtracting Decimals ...................................................31
   Multiplying and Dividing Decimals ................................................32
   Comparing Decimals .......................................................................33
   Rounding Decimals .........................................................................34
   Answers of Worksheets ...................................................................35

**Chapter 3 : Proportions, Ratios, and Percent ..................................38**
   Simplifying Ratios ...........................................................................39
   Proportional Ratios ..........................................................................40
   Similarity and Ratios .......................................................................41
   Ratio and Rates Word Problems ......................................................42
   Percentage Calculations ...................................................................43
   Percent Problems .............................................................................44
   Discount, Tax and Tip .....................................................................45
   Percent of Change ...........................................................................46
   Simple Interest .................................................................................47

WWW.MathNotion.Com

# ACT Aspire Subject Test Mathematics Grade 8

    Answers of Worksheets ................................................................................................48

## Chapter 4 : Exponents and Radicals Expressions................................................51
    Multiplication Property of Exponents ..........................................................................52
    Zero and Negative Exponents ....................................................................................53
    Division Property of Exponents ..................................................................................54
    Powers of Products and Quotients .............................................................................55
    Negative Exponents and Negative Bases ..................................................................56
    Scientific Notation .......................................................................................................57
    Square Roots .............................................................................................................58
    Simplifying Radical Expressions .................................................................................59
    Multiplying Radical Expressions .................................................................................60
    Simplifying Radical Expressions Involving Fractions...................................................61
    Adding and Subtracting Radical Expressions .............................................................62
    Answers of Worksheets ..............................................................................................63

## Chapter 5 : Algebraic Expressions .......................................................................68
    Simplifying Variable Expressions ................................................................................69
    Simplifying Polynomial Expressions ...........................................................................70
    Translate Phrases into an Algebraic Statement .........................................................71
    The Distributive Property ............................................................................................72
    Evaluating One Variable Expressions .........................................................................73
    Evaluating Two Variables Expressions .......................................................................74
    Combining like Terms .................................................................................................75
    Answers of Worksheets ..............................................................................................76

## Chapter 6 Equations and Inequalities..................................................................78
    One–Step Equations ..................................................................................................79
    Multi–Step Equations ..................................................................................................80
    Graphing Single–Variable Inequalities .......................................................................81
    One–Step Inequalities ................................................................................................82
    Multi-Step Inequalities ................................................................................................83
    Systems of Equations ................................................................................................84
    Systems of Equations Word Problems .......................................................................85
    Answers of Worksheets ..............................................................................................86

## Chapter 7 : Linear Functions................................................................................89
    Finding Slope..............................................................................................................90
    Graphing Lines Using Line Equation ..........................................................................91
    Writing Linear Equations ............................................................................................92
    Graphing Linear Inequalities ......................................................................................93

WWW.MathNotion.Com

# ACT Aspire Subject Test Mathematics Grade 8

    Finding Midpoint ..................................................................................................................94

    Finding Distance of Two Points ...........................................................................................95

    Answers of Worksheets .......................................................................................................96

**Chapter 8 : Polynomials** ................................................................................................99

    Writing Polynomials in Standard Form .............................................................................100

    Simplifying Polynomials .....................................................................................................101

    Adding and Subtracting Polynomials ................................................................................102

    Multiplying Monomials .......................................................................................................103

    Multiplying and Dividing Monomials ..................................................................................104

    Multiplying a Polynomial and a Monomial ........................................................................105

    Multiplying Binomials .........................................................................................................106

    Factoring Trinomials ..........................................................................................................107

    Operations with Polynomials .............................................................................................108

    Answers of Worksheets .....................................................................................................109

**Chapter 9 : Functions Operations and Quadratic** ..................................................113

    Evaluating Function ...........................................................................................................114

    Adding and Subtracting Functions ....................................................................................115

    Multiplying and Dividing Functions ....................................................................................116

    Composition of Functions ..................................................................................................117

    Quadratic Equation ............................................................................................................118

    Solving Quadratic Equations .............................................................................................119

    Quadratic Formula and the Discriminant ..........................................................................120

    Graphing Quadratic Functions ..........................................................................................121

    Quadratic Inequalities ........................................................................................................122

    Domain and Range of Radical Functions .........................................................................123

    Solving Radical Equations .................................................................................................124

    Answers of Worksheets .....................................................................................................125

**Chapter 10 : Geometry and Solid Figures** ...............................................................130

    Angles ................................................................................................................................131

    Pythagorean Relationship .................................................................................................132

    Triangles ............................................................................................................................133

    Polygons ............................................................................................................................134

    Trapezoids .........................................................................................................................135

    Circles ................................................................................................................................136

    Cubes ................................................................................................................................137

    Rectangular Prism .............................................................................................................137

    Cylinder .............................................................................................................................139

WWW.MathNotion.Com

# ACT Aspire Subject Test Mathematics Grade 8

    Pyramids and Cone.................................................................................................140

    Answers of Worksheets..........................................................................................141

**Chapter 11 : Statistics and Probability ..................................................................143**

    Mean and Median ..................................................................................................144

    Mode and Range ...................................................................................................145

    Times Series ..........................................................................................................146

    Stem–and–Leaf Plot...............................................................................................147

    Pie Graph ..............................................................................................................148

    Probability Problems.............................................................................................149

    Factorials ..............................................................................................................150

    Combinations and Permutations ..........................................................................151

    Answers of Worksheets..........................................................................................152

**Chapter 12 : ACT Aspire Math Practice Tests .......................................................155**

    ACT Aspire Practice Test 1 ...................................................................................159

    ACT Aspire Practice Test 2 ...................................................................................177

**Chapter 13 : Answers and Explanations................................................................191**

    Answer Key............................................................................................................191

    Practice Test 1 ......................................................................................................193

    Practice Test 2 ......................................................................................................203

ACT Aspire Subject Test Mathematics Grade 8

# Chapter 1 :
# Integers and Number Theory

**Topics that you will practice in this chapter:**

- ✓ Rounding
- ✓ Whole Number Addition and Subtraction
- ✓ Whole Number Multiplication and Division
- ✓ Rounding and Estimates
- ✓ Adding and Subtracting Integers
- ✓ Multiplying and Dividing Integers
- ✓ Order of Operations
- ✓ Ordering Integers and Numbers
- ✓ Integers and Absolute Value
- ✓ Factoring Numbers
- ✓ Greatest Common Factor (GCF)
- ✓ Least Common Multiple (LCM)

*"Wherever there is number, there is beauty."* −Proclus

ACT Aspire Subject Test Mathematics Grade 8

# Rounding

✎ **Round each number to the nearest ten.**

1) 42 = ___
2) 88 = ___
3) 24 = ___
4) 57 = ___

5) 19 = ___
6) 25 = ___
7) 93 = ___
8) 71 = ___

9) 48 = ___
10) 81 = ___
11) 58 = ___
12) 87 = ___

✎ **Round each number to the nearest hundred.**

13) 198 = ___
14) 387 = ___
15) 816 = ___
16) 101 = ___

17) 321 = ___
18) 433 = ___
19) 579 = ___
20) 825 = ___

21) 580 = ___
22) 868 = ___
23) 480 = ___
24) 287 = ___

✎ **Round each number to the nearest thousand.**

25) 1,382 = ___
26) 3,420 = ___
27) 4,254 = ___
28) 6,861 = ___

29) 9,099 = ___
30) 22,980 = ___
31) 45,188 = ___
32) 16,808 = ___

33) 52,866 = ___
34) 85,190 = ___
35) 70,990 = ___
36) 26,869 = ___

WWW.MathNotion.Com

# ACT Aspire Subject Test Mathematics Grade 8

## Rounding and Estimates

✎ **Estimate the sum by rounding each number to the nearest ten.**

1) 13 + 22 = _____

2) 71 + 23 = _____

3) 61 + 58 = _____

4) 56 + 85 = _____

5) 368 + 249 = _____

6) 330 + 903 = _____

7) 471 + 293 = _____

8) 1,950 + 2,655 = _____

✎ **Estimate the product by rounding each number to the nearest ten.**

9) 32 × 71 = _____

10) 12 × 33 = _____

11) 31 × 83 = _____

12) 19 × 11 = _____

13) 42 × 76 = _____

14) 63 × 34 = _____

15) 19 × 31 = _____

16) 59 × 71 = _____

✎ **Estimate the sum or product by rounding each number to the nearest ten.**

17) $\begin{array}{r} 29 \\ \times\ 12 \\ \hline \end{array}$

18) $\begin{array}{r} 37 \\ \times\ 26 \\ \hline \end{array}$

19) $\begin{array}{r} 48 \\ +\ 82 \\ \hline \end{array}$

20) $\begin{array}{r} 65 \\ +44 \\ \hline \end{array}$

21) $\begin{array}{r} 37 \\ \times\ 14 \\ \hline \end{array}$

22) $\begin{array}{r} 71 \\ +\ 32 \\ \hline \end{array}$

WWW.MathNotion.Com

ACT Aspire Subject Test Mathematics Grade 8

## Adding and Subtracting Integers

✎ **Find each sum.**

1) $14 + (-6) =$

2) $(-13) + (-20) =$

3) $5 + (-28) =$

4) $50 + (-12) =$

5) $(-7) + (-15) + 3 =$

6) $30 + (-14) + 8 =$

7) $40 + (-10) + (-14) + 17 =$

8) $(-15) + (-20) + 13 + 35 =$

9) $40 + (-20) + (38 - 29) =$

10) $28 + (-12) + (30 - 12) =$

✎ **Find each difference.**

11) $(-18) - (-7) =$

12) $25 - (-14) =$

13) $(-20) - 36 =$

14) $34 - (-19) =$

15) $51 - (30 - 21) =$

16) $17 - (5) - (-24) =$

17) $(35 + 20) - (-46) =$

18) $48 - 16 - (-8) =$

19) $62 - (28 + 17) - (-15) =$

20) $58 - (-23) - (-31) =$

21) $19 - (-8) - (-13) =$

22) $(19 - 24) - (-14) =$

23) $27 - 33 - (-21) =$

24) $58 - (32 + 24) - (-9) =$

25) $36 - (-30) + (-17) =$

26) $27 - (-42) + (-31) =$

WWW.MathNotion.Com

ACT Aspire Subject Test Mathematics Grade 8

## Multiplying and Dividing Integers

✎ **Find each product.**

1) $(-9) \times (-5) =$

2) $(-3) \times 9 =$

3) $8 \times (-12) =$

4) $(-7) \times (-20) =$

5) $(-3) \times (-5) \times 6 =$

6) $(14 - 3) \times (-8) =$

7) $12 \times (-9) \times (-3) =$

8) $(140 + 10) \times (-2) =$

9) $10 \times (-12 + 8) \times 3 =$

10) $(-8) \times (-5) \times (-10) =$

✎ **Find each quotient.**

11) $42 \div (-7) =$

12) $(-48) \div (-6) =$

13) $(-40) \div (-8) =$

14) $54 \div (-2) =$

15) $152 \div 19 =$

16) $(-144) \div (-12) =$

17) $180 \div (-10) =$

18) $(-312) \div (-12) =$

19) $221 \div (-13) =$

20) $(-126) \div (6) =$

21) $(-161) \div (-7) =$

22) $-266 \div (-14) =$

23) $(-120) \div (-4) =$

24) $270 \div (-18) =$

25) $(-208) \div (-8) =$

26) $(135) \div (-15) =$

WWW.MathNotion.Com

# ACT Aspire Subject Test Mathematics Grade 8

## Order of Operations

✏ **Evaluate each expression.**

1) $7 + (5 \times 4) =$

2) $14 - (3 \times 6) =$

3) $(19 \times 4) + 16 =$

4) $(16 - 7) - (8 \times 2) =$

5) $27 + (18 \div 3) =$

6) $(18 \times 8) \div 6 =$

7) $(32 \div 4) \times (-2) =$

8) $(9 \times 4) + (32 - 18) =$

9) $24 + (4 \times 3) + 7 =$

10) $(36 \times 3) \div (2 + 2) =$

11) $(-7) + (12 \times 3) + 11 =$

12) $(8 \times 5) - (24 \div 6) =$

13) $(7 \times 6 \div 3) - (12 + 9) =$

14) $(13 + 5 - 14) \times 3 - 2 =$

15) $(20 - 14 + 30) \times (64 \div 4) =$

16) $32 + (28 - (36 \div 9)) =$

17) $(7 + 6 - 4 - 7) + (15 \div 5) =$

18) $(85 - 20) + (20 - 18 + 7) =$

19) $(20 \times 2) + (14 \times 3) - 22 =$

20) $18 + 5 - (30 \times 3) + 20 =$

# ACT Aspire Subject Test Mathematics Grade 8

## Ordering Integers and Numbers

✎ **Order each set of integers from least to greatest.**

1) $8, -10, -5, -3, 4$      ___, ___, ___, ___, ___, ___

2) $-10, -18, 6, 14, 27$      ___, ___, ___, ___, ___, ___

3) $15, -8, -21, 21, -23$      ___, ___, ___, ___, ___, ___

4) $-14, -40, 23, -12, 47$      ___, ___, ___, ___, ___, ___

5) $59, -54, 32, -57, 36$      ___, ___, ___, ___, ___, ___

6) $68, 26, -19, 47, -34$      ___, ___, ___, ___, ___, ___

✎ **Order each set of integers from greatest to least.**

7) $18, 36, -16, -18, -10$      ___, ___, ___, ___, ___, ___

8) $27, 34, -12, -24, 94$      ___, ___, ___, ___, ___, ___

9) $50, -21, -13, 42, -2$      ___, ___, ___, ___, ___, ___

10) $37, 46, -20, -16, 86$      ___, ___, ___, ___, ___, ___

11) $-18, 88, -26, -59, 75$      ___, ___, ___, ___, ___, ___

12) $-65, -30, -25, 3, 14$      ___, ___, ___, ___, ___, ___

WWW.MathNotion.Com

ACT Aspire Subject Test Mathematics Grade 8

## Integers and Absolute Value

✎ **Write absolute value of each number.**

1) $|-2| =$

2) $|-27| =$

3) $|-20| =$

4) $|14| =$

5) $|6| =$

6) $|-55| =$

7) $|16| =$

8) $|2| =$

9) $|54| =$

10) $|-4| =$

11) $|-11|$

12) $|88| =$

13) $|0| =$

14) $|79| =$

15) $|-32| =$

16) $|-17| =$

17) $|42| =$

18) $|-46| =$

19) $|1| =$

20) $|-40| =$

✎ **Evaluate the value.**

21) $|-5| - \frac{|-21|}{7} =$

22) $14 - |3 - 15| - |-4| =$

23) $\frac{|-32|}{4} \times |-4| =$

24) $\frac{|7 \times (-3)|}{7} \times \frac{|-19|}{3} =$

25) $|4 \times (-5)| + \frac{|-40|}{5} =$

26) $\frac{|-45|}{9} \times \frac{|-24|}{12} =$

27) $|-12 + 8| \times \frac{|-7 \times 7|}{7} =$

28) $\frac{|-11 \times 2|}{4} \times |-16| =$

WWW.MathNotion.Com

ACT Aspire Subject Test Mathematics Grade 8

## Factoring Numbers

✏️ **List all positive factors of each number.**

1) 9

2) 16

3) 24

4) 30

5) 26

6) 46

7) 20

8) 68

9) 28

10) 98

11) 14

12) 54

13) 55

14) 18

15) 63

16) 34

17) 50

18) 62

19) 95

20) 64

21) 70

22) 45

23) 22

24) 65

ACT Aspire Subject Test Mathematics Grade 8

## Greatest Common Factor

✏ **Find the GCF for each number pair.**

1) 6, 2

2) 4, 5

3) 3, 12

4) 7, 3

5) 5, 10

6) 8, 48

7) 6, 18

8) 9, 15

9) 12, 18

10) 4, 36

11) 6, 10

12) 28, 52

13) 25, 10

14) 22, 24

15) 9, 54

16) 8, 54

17) 42, 14

18) 16, 40

19) 9, 2, 3

20) 5, 15, 10

21) 7, 9, 2

22) 16, 64

23) 30, 48

24) 36, 63

ACT Aspire Subject Test Mathematics Grade 8

## Least Common Multiple

✎ **Find the LCM for each number pair.**

1) 6, 9

2) 15, 45

3) 16, 40

4) 12, 36

5) 18, 27

6) 14, 42

7) 6, 30

8) 8, 56

9) 7, 21

10) 8, 20

11) 15, 25

12) 7, 9

13) 4, 11

14) 8, 28

15) 28, 56

16) 40, 50

17) 12, 13

18) 22, 11

19) 36, 20

20) 15, 35

21) 18, 81

22) 30, 54

23) 18, 45

24) 75, 25

ACT Aspire Subject Test Mathematics Grade 8

# Answers of Worksheets

**Rounding**

| | | | |
|---|---|---|---|
| 1) 40 | 10) 80 | 19) 600 | 28) 7,000 |
| 2) 90 | 11) 60 | 20) 800 | 29) 9,000 |
| 3) 20 | 12) 90 | 21) 600 | 30) 23,000 |
| 4) 60 | 13) 200 | 22) 900 | 31) 45,000 |
| 5) 20 | 14) 400 | 23) 500 | 32) 17,000 |
| 6) 30 | 15) 800 | 24) 300 | 33) 53,000 |
| 7) 90 | 16) 100 | 25) 1,000 | 34) 85,000 |
| 8) 70 | 17) 300 | 26) 3,000 | 35) 71,000 |
| 9) 50 | 18) 400 | 27) 4,000 | 36) 27,000 |

**Rounding and Estimates**

| | | | |
|---|---|---|---|
| 1) 30 | 7) 760 | 13) 3,200 | 19) 130 |
| 2) 90 | 8) 4,610 | 14) 1,800 | 20) 110 |
| 3) 120 | 9) 2,100 | 15) 600 | 21) 400 |
| 4) 150 | 10) 300 | 16) 4,200 | 22) 100 |
| 5) 620 | 11) 2,400 | 17) 300 | |
| 6) 1,230 | 12) 200 | 18) 1,200 | |

**Adding and Subtracting Integers**

| | | | |
|---|---|---|---|
| 1) 8 | 8) 13 | 15) 42 | 22) 9 |
| 2) −33 | 9) 29 | 16) 36 | 23) 15 |
| 3) −23 | 10) 34 | 17) 101 | 24) 11 |
| 4) 38 | 11) −11 | 18) 40 | 25) 49 |
| 5) −19 | 12) 39 | 19) 32 | 26) 38 |
| 6) 24 | 13) −56 | 20) 112 | |
| 7) 33 | 14) 53 | 21) 40 | |

**Multiplying and Dividing Integers**

| | | | |
|---|---|---|---|
| 1) 45 | 6) −88 | 11) −6 | 16) 12 |
| 2) −27 | 7) 324 | 12) 8 | 17) −18 |
| 3) −96 | 8) −300 | 13) 5 | 18) 26 |
| 4) 140 | 9) −120 | 14) −27 | 19) −17 |
| 5) 90 | 10) −400 | 15) 8 | 20) −21 |

WWW.MathNotion.Com

# ACT Aspire Subject Test Mathematics Grade 8

21) 23
22) 19
23) 30
24) −15
25) 26
26) −9

**Order of Operations**

1) 27
2) −4
3) 92
4) −7
5) 33
6) 24
7) −16
8) 50
9) 43
10) 27
11) 40
12) 36
13) −7
14) 10
15) 576
16) 56
17) 5
18) 74
19) 60
20) −47

**Ordering Integers and Numbers**

1) −10, −5, −3, 4, 8
2) −18, −10, 6, 14, 27
3) −23, −21, −8, 15, 21
4) −40, −14, −12, 23, 47
5) −57, −54, 32, 36, 59
6) −34, −19, 26, 47, 68
7) 36, 18, −10, −16, −18
8) 94, 34, 27, −12, −24
9) 50, 42, −2, −13, −21
10) 86, 46, 37, −16, −20
11) 88, 75, −18, −26, −59
12) 14, 3, −25, −30, −65

**Integers and Absolute Value**

1) 2
2) 27
3) 20
4) 14
5) 6
6) 55
7) 16
8) 2
9) 54
10) 4
11) 11
12) 88
13) 0
14) 79
15) 32
16) 17
17) 42
18) 46
19) 1
20) 40
21) 2
22) −2
23) 32
24) 19
25) 28
26) 10
27) 28
28) 88

**Factoring Numbers**

1) 1, 3, 9
2) 1, 2, 4, 8, 16
3) 1, 2, 3, 4, 6, 8, 12, 24
4) 1, 2, 3, 5, 6, 10, 15, 30
5) 1, 2, 13, 26
6) 1, 2, 23, 46
7) 1, 2, 4, 5, 10, 20
8) 1, 2, 4, 17, 34, 68
9) 1, 2, 4, 7, 14, 28
10) 1, 2, 7, 14, 49, 98
11) 1, 2, 7, 14
12) 1, 2, 3, 6, 9, 18, 27, 54
13) 1, 5, 11, 55
14) 1, 2, 3, 6, 9, 18
15) 1, 3, 7, 9, 21, 63
16) 1, 2, 17, 34
17) 1, 2, 5, 10, 25, 50
18) 1, 2, 31, 62
19) 1, 5, 19, 95
20) 1, 2, 4, 8, 16, 32, 64
21) 1, 2, 5, 7, 10, 14, 35, 70
22) 1, 3, 5, 9, 15, 45
23) 1, 2, 11, 22
24) 1, 5, 13, 65

WWW.MathNotion.Com

# ACT Aspire Subject Test Mathematics Grade 8

**Greatest Common Factor**

1) 2
2) 1
3) 3
4) 1
5) 5
6) 8
7) 6
8) 3
9) 6
10) 4
11) 2
12) 4
13) 5
14) 2
15) 9
16) 2
17) 14
18) 8
19) 1
20) 5
21) 1
22) 16
23) 6
24) 9

**Least Common Multiple**

1) 18
2) 45
3) 80
4) 36
5) 54
6) 42
7) 30
8) 56
9) 21
10) 40
11) 75
12) 63
13) 44
14) 56
15) 56
16) 200
17) 156
18) 22
19) 180
20) 105
21) 162
22) 270
23) 90
24) 75

ACT Aspire Subject Test Mathematics Grade 8

# Chapter 2 :
# Fractions and Decimals

**Topics that you will practice in this chapter:**

- ✓ Simplifying Fractions
- ✓ Adding and Subtracting Fractions
- ✓ Multiplying and Dividing Fractions
- ✓ Adding and Subtract Mixed Numbers
- ✓ Multiplying and Dividing Mixed Numbers
- ✓ Adding and Subtracting Decimals
- ✓ Multiplying and Dividing Decimals
- ✓ Comparing Decimals
- ✓ Rounding Decimals

*"A Man is like a fraction whose numerator is what he is and whose denominator is what he thinks of himself. The larger the denominator, the smaller the fraction."* −Tolstoy

# ACT Aspire Subject Test Mathematics Grade 8

## Simplifying Fractions

✎ Simplify each fraction to its lowest terms.

1) $\dfrac{5}{10} =$

2) $\dfrac{28}{35} =$

3) $\dfrac{27}{36} =$

4) $\dfrac{40}{80} =$

5) $\dfrac{14}{56} =$

6) $\dfrac{32}{48} =$

7) $\dfrac{52}{65} =$

8) $\dfrac{15}{60} =$

9) $\dfrac{80}{160} =$

10) $\dfrac{55}{77} =$

11) $\dfrac{28}{112} =$

12) $\dfrac{32}{64} =$

13) $\dfrac{63}{72} =$

14) $\dfrac{81}{90} =$

15) $\dfrac{35}{105} =$

16) $\dfrac{25}{70} =$

17) $\dfrac{80}{280} =$

18) $\dfrac{12}{81} =$

19) $\dfrac{36}{186} =$

20) $\dfrac{240}{540} =$

21) $\dfrac{70}{560} =$

✎ Find the answer for each problem.

22) Which of the following fractions equal to $\dfrac{3}{4}$? ____

   A. $\dfrac{60}{90}$   B. $\dfrac{43}{104}$   C. $\dfrac{48}{64}$   D. $\dfrac{150}{300}$

23) Which of the following fractions equal to $\dfrac{5}{8}$? ____

   A. $\dfrac{125}{200}$   B. $\dfrac{115}{200}$   C. $\dfrac{50}{100}$   D. $\dfrac{30}{90}$

24) Which of the following fractions equal to $\dfrac{3}{7}$? ____

   A. $\dfrac{58}{116}$   B. $\dfrac{54}{126}$   C. $\dfrac{270}{167}$   D. $\dfrac{42}{63}$

# ACT Aspire Subject Test Mathematics Grade 8

## Adding and Subtracting Fractions

✏️ **Find the sum.**

1) $\frac{5}{9} + \frac{4}{9} =$

2) $\frac{1}{2} + \frac{1}{7} =$

3) $\frac{3}{8} + \frac{1}{4} =$

4) $\frac{3}{5} + \frac{1}{2} =$

5) $\frac{1}{4} + \frac{3}{5} =$

6) $\frac{7}{8} + \frac{3}{8} =$

7) $\frac{1}{2} + \frac{7}{10} =$

8) $\frac{2}{5} + \frac{2}{3} =$

9) $\frac{5}{7} + \frac{2}{3} =$

10) $\frac{7}{12} + \frac{3}{4} =$

11) $\frac{5}{6} + \frac{2}{5} =$

12) $\frac{1}{12} + \frac{2}{3} =$

✏️ **Find the difference.**

13) $\frac{1}{3} - \frac{1}{6} =$

14) $\frac{3}{4} - \frac{1}{8} =$

15) $\frac{1}{2} - \frac{1}{3} =$

16) $\frac{1}{4} - \frac{1}{5} =$

17) $\frac{5}{8} - \frac{2}{3} =$

18) $\frac{1}{4} - \frac{1}{7} =$

19) $\frac{5}{6} - \frac{1}{9} =$

20) $\frac{3}{4} - \frac{1}{6} =$

21) $\frac{7}{8} - \frac{1}{12} =$

22) $\frac{8}{15} - \frac{3}{5} =$

23) $\frac{3}{12} - \frac{1}{14} =$

24) $\frac{10}{13} - \frac{7}{26} =$

25) $\frac{6}{7} - \frac{3}{4} =$

26) $\frac{4}{5} - \frac{1}{8} =$

27) $\frac{4}{7} - \frac{2}{35} =$

28) $\frac{9}{16} - \frac{2}{8} =$

29) $\frac{8}{9} - \frac{7}{18} =$

30) $\frac{1}{2} - \frac{4}{9} =$

ACT Aspire Subject Test Mathematics Grade 8

## Multiplying and Dividing Fractions

✏️ Find the value of each expression in lowest terms.

1) $\frac{1}{5} \times \frac{15}{5} =$

2) $\frac{9}{12} \times \frac{4}{9} =$

3) $\frac{1}{16} \times \frac{8}{10} =$

4) $\frac{1}{24} \times \frac{8}{10} =$

5) $\frac{1}{5} \times \frac{1}{4} =$

6) $\frac{7}{9} \times \frac{1}{7} =$

7) $\frac{6}{7} \times \frac{1}{3} =$

8) $\frac{2}{8} \times \frac{2}{8} =$

9) $\frac{5}{8} \times \frac{3}{5} =$

10) $\frac{4}{7} \times \frac{1}{8} =$

11) $\frac{7}{15} \times \frac{5}{7} =$

12) $\frac{3}{10} \times \frac{5}{9} =$

✏️ Find the value of each expression in lowest terms.

13) $\frac{1}{4} \div \frac{1}{8} =$

14) $\frac{1}{10} \div \frac{1}{5} =$

15) $\frac{3}{4} \div \frac{1}{5} =$

16) $\frac{1}{3} \div \frac{5}{6} =$

17) $\frac{1}{7} \div \frac{8}{42} =$

18) $\frac{3}{4} \div \frac{1}{6} =$

19) $\frac{2}{7} \div \frac{7}{13} =$

20) $\frac{1}{24} \div \frac{3}{16} =$

21) $\frac{7}{12} \div \frac{5}{6} =$

22) $\frac{22}{18} \div \frac{11}{9} =$

23) $\frac{9}{35} \div \frac{3}{7} =$

24) $\frac{2}{7} \div \frac{8}{21} =$

25) $\frac{1}{9} \div \frac{2}{5} =$

26) $\frac{5}{12} \div \frac{3}{5} =$

27) $\frac{3}{20} \div \frac{1}{6} =$

28) $\frac{8}{20} \div \frac{3}{4} =$

29) $\frac{5}{6} \div \frac{2}{9} =$

30) $\frac{5}{11} \div \frac{3}{4} =$

WWW.MathNotion.Com

ACT Aspire Subject Test Mathematics Grade 8

## Adding and Subtracting Mixed Numbers

✏ **Find the sum.**

1) $3\frac{1}{3} + 2\frac{1}{6} =$

2) $4\frac{1}{2} + 3\frac{1}{2} =$

3) $3\frac{3}{8} + 1\frac{1}{8} =$

4) $2\frac{1}{4} + 2\frac{1}{3} =$

5) $3\frac{5}{6} + 2\frac{7}{12} =$

6) $5\frac{4}{15} + 3\frac{3}{5} =$

7) $2\frac{1}{3} + 4\frac{3}{7} =$

8) $3\frac{1}{2} + 4\frac{2}{5} =$

9) $5\frac{2}{5} + 6\frac{3}{7} =$

10) $8\frac{5}{16} + 6\frac{1}{12} =$

✏ **Find the difference.**

11) $3\frac{1}{4} - 1\frac{3}{4} =$

12) $6\frac{3}{5} - 4\frac{2}{5} =$

13) $4\frac{1}{3} - 3\frac{1}{9} =$

14) $7\frac{1}{7} - 5\frac{1}{2} =$

15) $5\frac{1}{3} - 2\frac{1}{12} =$

16) $8\frac{1}{5} - 4\frac{1}{3} =$

17) $9\frac{1}{4} - 6\frac{1}{8} =$

18) $11\frac{7}{15} - 8\frac{3}{5} =$

19) $14\frac{5}{6} - 11\frac{3}{5} =$

20) $18\frac{2}{7} - 14\frac{1}{5} =$

21) $9\frac{1}{3} - 4\frac{1}{4} =$

22) $6\frac{1}{8} - 4\frac{1}{16} =$

23) $19\frac{3}{8} - 15\frac{1}{3} =$

24) $11\frac{1}{9} - 8\frac{1}{8} =$

25) $17\frac{1}{7} - 11\frac{1}{5} =$

26) $16\frac{2}{9} - 9\frac{5}{7} =$

# ACT Aspire Subject Test Mathematics Grade 8

## Multiplying and Dividing Mixed Numbers

✎ **Find the product.**

1) $5\frac{1}{2} \times 2\frac{1}{4} =$

2) $5\frac{1}{3} \times 4\frac{1}{3} =$

3) $5\frac{3}{4} \times 6\frac{1}{4} =$

4) $3\frac{1}{3} \times 2\frac{3}{5} =$

5) $4\frac{8}{10} \times 1\frac{1}{24} =$

6) $6\frac{2}{7} \times 1\frac{1}{11} =$

7) $8\frac{2}{3} \times 3\frac{1}{2} =$

8) $3\frac{4}{7} \times 2\frac{1}{5} =$

9) $5\frac{2}{8} \times 4\frac{1}{6} =$

10) $7\frac{3}{3} \times 1\frac{3}{8} =$

✎ **Find the quotient.**

11) $2\frac{2}{5} \div 4\frac{1}{5} =$

12) $4\frac{1}{6} \div 3\frac{1}{3} =$

13) $6\frac{1}{3} \div 1\frac{1}{2} =$

14) $7\frac{1}{10} \div 2\frac{2}{5} =$

15) $3\frac{1}{3} \div 1\frac{1}{9} =$

16) $1\frac{1}{10} \div 4\frac{1}{2} =$

17) $1\frac{3}{16} \div 5\frac{1}{4} =$

18) $4\frac{1}{3} \div 4\frac{3}{4} =$

19) $9\frac{1}{3} \div 2\frac{1}{4} =$

20) $15\frac{1}{3} \div 5\frac{1}{2} =$

21) $4\frac{1}{6} \div 1\frac{1}{5} =$

22) $1\frac{1}{18} \div 1\frac{2}{9} =$

23) $4\frac{2}{7} \div 1\frac{3}{10} =$

24) $7\frac{1}{3} \div 2\frac{2}{11} =$

25) $8\frac{2}{5} \div 1\frac{1}{6} =$

26) $9\frac{1}{3} \div 2\frac{1}{7} =$

WWW.MathNotion.Com

# ACT Aspire Subject Test Mathematics Grade 8

## Adding and Subtracting Decimals

✎ **Add and subtract decimals.**

1) 35.19 − 24.28 = _____

2) 34.29 + 42.58 = _____

3) 61.20 + 33.75 = _____

4) 38.72 − 21.68 = _____

5) 57.39 + 26.54 = _____

6) 70.24 − 42.35 = _____

7) 86.09 − 35.14 = _____

8) 54.51 + 32.66 = _____

9) 114.21 − 88.69 = _____

✎ **Find the missing number.**

10) ___ + 2.8 = 5.4

11) 4.1 + ___ = 5.88

12) 6.45 + ___ = 8

13) 7.25 − ___ = 3.40

14) ___ − 2.35 = 4.25

15) ___ − 19.85 = 6.54

16) 22.15 + ___ = 28.95

17) ___ − 37.16 = 9.42

18) ___ + 24.50 = 34.19

19) 72.40 + ___ = 125.20

WWW.MathNotion.Com

# ACT Aspire Subject Test Mathematics Grade 8

## Multiplying and Dividing Decimals

✏ **Find the product.**

1) $0.5 \times 0.6 =$

2) $3.3 \times 0.4 =$

3) $1.28 \times 0.5 =$

4) $0.35 \times 0.6 =$

5) $1.85 \times 0.6 =$

6) $0.24 \times 0.5 =$

7) $5.25 \times 1.4 =$

8) $18.5 \times 4.6 =$

9) $15.4 \times 6.8 =$

10) $19.5 \times 2.6 =$

11) $32.2 \times 1.5 =$

12) $78.4 \times 4.5 =$

✏ **Find the quotient.**

13) $1.85 \div 10 =$

14) $74.6 \div 100 =$

15) $3.6 \div 3 =$

16) $9.6 \div 0.4 =$

17) $15.5 \div 0.5 =$

18) $32.8 \div 0.2 =$

19) $22.15 \div 1,000 =$

20) $53.55 \div 0.7 =$

21) $322.2 \div 0.2 =$

22) $50.67 \div 0.18 =$

23) $77.4 \div 0.8 =$

24) $27.93 \div 0.03 =$

# ACT Aspire Subject Test Mathematics Grade 8

## Comparing Decimals

✏️ **Write the correct comparison symbol (>, < or =).**

1) 0.70 ☐ 0.070

2) 0.049 ☐ 0.49

3) 5.090 ☐ 5.09

4) 2.57 ☐ 2.05

5) 9.03 ☐ 0.930

6) 6.06 ☐ 6.6

7) 7.02 ☐ 7.020

8) 3.04 ☐ 3.2

9) 3.61 ☐ 3.245

10) 0.986 ☐ 0.0986

11) 17.24 ☐ 17.240

12) 0.759 ☐ 0.81

13) 9.040 ☐ 9.40

14) 5.73 ☐ 5.213

15) 9.44 ☐ 9.404

16) 7.17 ☐ 7.170

17) 4.85 ☐ 4.085

18) 9.041 ☐ 9.40

19) 3.033 ☐ 3.030

20) 4.97 ☐ 4.970

# Rounding Decimals

🖎 **Round each decimal to the nearest whole number.**

1) 28.12           3) 16.22           5) 7.95

2) 6.9             4) 8.5             6) 52.7

🖎 **Round each decimal to the nearest tenth.**

7) 31.761          9) 94.729          11) 13.219

8) 14.421          10) 77.89          12) 59.89

🖎 **Round each decimal to the nearest hundredth.**

13) 8.428          15) 55.3786        17) 62.241

14) 23.812         16) 231.912        18) 19.447

🖎 **Round each decimal to the nearest thousandth.**

19) 15.54324       21) 243.8652       23) 67.1983

20) 34.62586       22) 80.4529        24) 72.36788

# ACT Aspire Subject Test Mathematics Grade 8

## Answers of Worksheets

**Simplifying Fractions**

1) $\frac{1}{2}$
2) $\frac{4}{5}$
3) $\frac{3}{4}$
4) $\frac{1}{2}$
5) $\frac{1}{4}$
6) $\frac{2}{3}$
7) $\frac{4}{5}$
8) $\frac{1}{4}$
9) $\frac{1}{2}$
10) $\frac{5}{7}$
11) $\frac{1}{4}$
12) $\frac{1}{2}$
13) $\frac{7}{8}$
14) $\frac{9}{10}$
15) $\frac{1}{3}$
16) $\frac{5}{14}$
17) $\frac{2}{7}$
18) $\frac{4}{27}$
19) $\frac{6}{31}$
20) $\frac{4}{9}$
21) $\frac{1}{8}$
22) C
23) A
24) B

**Adding and Subtracting Fractions**

1) $\frac{9}{9} = 1$
2) $\frac{9}{14}$
3) $\frac{5}{8}$
4) $1\frac{1}{10}$
5) $\frac{17}{20}$
6) $1\frac{1}{4}$
7) $1\frac{1}{5}$
8) $1\frac{1}{15}$
9) $1\frac{8}{21}$
10) $1\frac{1}{3}$
11) $1\frac{7}{30}$
12) $\frac{3}{4}$
13) $\frac{1}{6}$
14) $\frac{5}{8}$
15) $\frac{1}{6}$
16) $\frac{1}{20}$
17) $-\frac{1}{24}$
18) $\frac{3}{28}$
19) $\frac{13}{18}$
20) $\frac{7}{12}$
21) $\frac{19}{24}$
22) $-\frac{1}{15}$
23) $\frac{5}{28}$
24) $\frac{1}{2}$
25) $\frac{3}{28}$
26) $\frac{27}{40}$
27) $\frac{18}{35}$
28) $\frac{5}{16}$
29) $\frac{1}{2}$
30) $\frac{1}{18}$

**Multiplying and Dividing Fractions**

1) $\frac{3}{5}$
2) $\frac{1}{3}$
3) $\frac{1}{20}$
4) $\frac{1}{30}$
5) $\frac{1}{20}$
6) $\frac{1}{9}$
7) $\frac{2}{7}$
8) $\frac{1}{16}$
9) $\frac{3}{8}$
10) $\frac{1}{14}$
11) $\frac{1}{3}$
12) $\frac{1}{6}$
13) 2
14) $\frac{1}{2}$
15) $3\frac{3}{4}$
16) $\frac{2}{5}$

www.MathNotion.Com

## ACT Aspire Subject Test Mathematics Grade 8

17) $\frac{3}{4}$
18) $4\frac{1}{2}$
19) $\frac{26}{49}$
20) $\frac{2}{9}$

21) $\frac{7}{10}$
22) $1$
23) $\frac{3}{5}$
24) $\frac{3}{4}$

25) $\frac{5}{18}$
26) $\frac{25}{36}$
27) $\frac{9}{10}$
28) $\frac{8}{15}$

29) $3\frac{3}{4}$
30) $\frac{20}{33}$

### Adding and Subtracting Mixed Numbers

1) $5\frac{1}{2}$
2) $8$
3) $4\frac{1}{2}$
4) $4\frac{7}{12}$
5) $6\frac{5}{12}$
6) $8\frac{13}{15}$
7) $6\frac{16}{21}$

8) $7\frac{9}{10}$
9) $11\frac{29}{35}$
10) $14\frac{19}{48}$
11) $1\frac{1}{2}$
12) $2\frac{1}{5}$
13) $1\frac{2}{9}$
14) $1\frac{9}{14}$

15) $3\frac{1}{4}$
16) $3\frac{13}{15}$
17) $3\frac{1}{8}$
18) $2\frac{13}{15}$
19) $3\frac{7}{30}$
20) $4\frac{3}{35}$
21) $5\frac{1}{12}$

22) $2\frac{1}{16}$
23) $4\frac{1}{24}$
24) $2\frac{71}{72}$
25) $5\frac{33}{35}$
26) $6\frac{32}{63}$

### Multiplying and Dividing Mixed Numbers

1) $12\frac{3}{8}$
2) $23\frac{1}{9}$
3) $35\frac{15}{16}$
4) $8\frac{2}{3}$
5) $5$
6) $6\frac{6}{7}$
7) $30\frac{1}{3}$
8) $7\frac{6}{7}$
9) $21\frac{7}{8}$

10) $11$
11) $\frac{4}{7}$
12) $1\frac{1}{4}$
13) $4\frac{2}{9}$
14) $2\frac{23}{24}$
15) $3$
16) $\frac{11}{45}$
17) $\frac{19}{84}$
18) $\frac{52}{57}$

19) $4\frac{4}{27}$
20) $2\frac{26}{33}$
21) $3\frac{17}{36}$
22) $\frac{19}{22}$
23) $3\frac{27}{91}$
24) $3\frac{13}{36}$
25) $7\frac{1}{5}$
26) $4\frac{16}{45}$

### Adding and Subtracting Decimals

1) 10.91
2) 76.87
3) 94.95
4) 17.04

# ACT Aspire Subject Test Mathematics Grade 8

5) 83.93
6) 27.89
7) 50.95
8) 87.17
9) 25.52
10) 2.6
11) 1.78
12) 1.55
13) 3.85
14) 6.6
15) 26.39
16) 6.8
17) 46.58
18) 9.69
19) 52.8

**Multiplying and Dividing Decimals**

1) 0.3
2) 1.32
3) 0.64
4) 0.21
5) 1.11
6) 0.12
7) 7.35
8) 85.1
9) 104.72
10) 50.7
11) 48.3
12) 352.8
13) 0.185
14) 0.746
15) 1.2
16) 24
17) 31
18) 164
19) 0.02215
20) 76.5
21) 1,611
22) 281.5
23) 96.75
24) 931

**Comparing Decimals**

1) >
2) <
3) =
4) >
5) >
6) <
7) =
8) <
9) >
10) >
11) =
12) <
13) <
14) >
15) >
16) =
17) >
18) <
19) >
20) =

**Rounding Decimals**

1) 28
2) 7
3) 16
4) 9
5) 8
6) 53
7) 31.8
8) 14.4
9) 94.7
10) 77.9
11) 13.2
12) 59.9
13) 8.43
14) 23.81
15) 55.38
16) 231.91
17) 62.24
18) 19.45
19) 15.543
20) 34.626
21) 243.865
22) 80.453
23) 67.198
24) 72.368

ACT Aspire Subject Test Mathematics Grade 8

# Chapter 3 :
# Proportions, Ratios, and Percent

**Topics that you will practice in this chapter:**

- ✓ Simplifying Ratios
- ✓ Proportional Ratios
- ✓ Similarity and Ratios
- ✓ Ratio and Rates Word Problems
- ✓ Percentage Calculations
- ✓ Percent Problems
- ✓ Discount, Tax and Tip
- ✓ Percent of Change
- ✓ Simple Interest

*Without mathematics, there's nothing you can do. Everything around you is mathematics. Everything around you is numbers." – Shakuntala Devi*

# ACT Aspire Subject Test Mathematics Grade 8

## Simplifying Ratios

✎ **Reduce each ratio.**

1) 15: 20 = ___ : ___      9) 8: 48 = ___ : ___       17) 56: 72 = ___ : ___

2) 7: 70 = ___ : ___       10) 49: 63 = ___ : ___     18) 26: 13 = ___ : ___

3) 16: 28 = ___ : ___      11) 18: 27 = ___ : ___     19) 15: 45 = ___ : ___

4) 7: 21 = ___ : ___       12) 35: 10 = ___ : ___     20) 28: 4 = ___ : ___

5) 4: 40 = ___ : ___       13) 90: 9 = ___ : ___      21) 24: 48 = ___ : ___

6) 6: 48 = ___ : ___       14) 24: 32 = ___ : ___     22) 30: 24 = ___ : ___

7) 16: 64 = ___ : ___      15) 7: 56 = ___ : ___      23) 70: 140 = ___ : ___

8) 10: 25 = ___ : ___      16) 45: 63 = ___ : ___     24) 6: 180 = ___ : ___

✎ **Write each ratio as a fraction in simplest form.**

25) 6: 12 =          32) 7: 35 =          39) 20: 300 =

26) 30: 50 =         33) 40: 96 =         40) 30: 120 =

27) 15: 35 =         34) 12: 54 =         41) 56: 42 =

28) 9: 27 =          35) 44: 52 =         42) 26: 130 =

29) 8: 24 =          36) 12: 27 =         43) 66: 123 =

30) 18: 84 =         37) 15: 180 =        44) 70: 630 =

31) 7: 14 =          38) 39: 143 =        45) 75: 125 =

# ACT Aspire Subject Test Mathematics Grade 8

## Proportional Ratios

✎ Fill in the blanks; Calculate each proportion.

1) $3:8 = \_\_ : 48$
2) $2:5 = 20:\_\_$
3) $1:9 = \_\_ : 81$
4) $6:7 = 12:\_\_$
5) $9:2 = 63:\_\_$
6) $8:7 = \_\_ : 49$

7) $20:3 = \_\_ : 15$
8) $1:3 = \_\_ : 75$
9) $7:6 = \_\_ : 60$
10) $8:5 = \_\_ : 45$
11) $3:10 = 60:\_\_$
12) $6:11 = 42:\_\_$

✎ State if each pair of ratios form a proportion.

13) $\frac{3}{20}$ and $\frac{9}{60}$
14) $\frac{1}{7}$ and $\frac{6}{42}$
15) $\frac{3}{7}$ and $\frac{24}{56}$
16) $\frac{4}{9}$ and $\frac{12}{18}$

17) $\frac{1}{9}$ and $\frac{12}{81}$
18) $\frac{7}{8}$ and $\frac{21}{28}$
19) $\frac{9}{13}$ and $\frac{27}{39}$
20) $\frac{1}{8}$ and $\frac{8}{64}$

21) $\frac{6}{19}$ and $\frac{30}{85}$
22) $\frac{5}{9}$ and $\frac{40}{81}$
23) $\frac{9}{14}$ and $\frac{108}{168}$
24) $\frac{15}{23}$ and $\frac{360}{552}$

✎ Calculate each proportion.

25) $\frac{20}{25} = \frac{32}{x}$, $x = \_\_\_$
26) $\frac{1}{8} = \frac{32}{x}$, $x = \_\_\_$
27) $\frac{15}{5} = \frac{21}{x}$, $x = \_\_\_$
28) $\frac{1}{7} = \frac{x}{294}$, $x = \_\_\_$

29) $\frac{7}{9} = \frac{x}{81}$, $x = \_\_\_$
30) $\frac{1}{5} = \frac{13}{x}$, $x = \_\_\_$
31) $\frac{9}{5} = \frac{36}{x}$, $x = \_\_\_$
32) $\frac{6}{13} = \frac{48}{x}$, $x = \_\_\_$

33) $\frac{5}{8} = \frac{x}{88}$, $x = \_\_\_$
34) $\frac{4}{15} = \frac{x}{240}$, $x = \_\_\_$
35) $\frac{9}{19} = \frac{x}{266}$, $x = \_\_\_$
36) $\frac{7}{15} = \frac{x}{270}$, $x = \_\_\_$

# Similarity and Ratios

✏️ **Each pair of figures is similar. Find the missing side.**

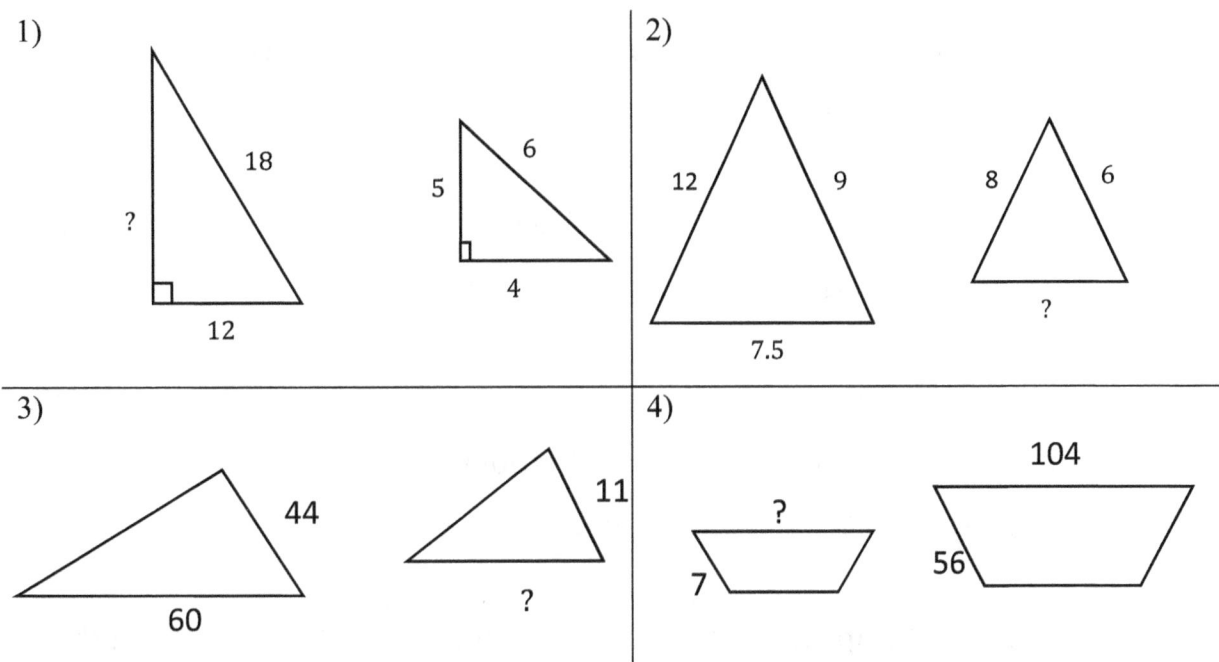

✏️ **Calculate.**

5) Two rectangles are similar. The first is 24 feet wide and 120 feet long. The second is 30 feet wide. What is the length of the second rectangle?
_____

6) Two rectangles are similar. One is 5 meters by 36 meters. The longer side of the second rectangle is 90 meters. What is the other side of the second rectangle? _____

7) A building casts a shadow 25 ft long. At the same time a girl 10 ft tall casts a shadow 5 ft long. How tall is the building? _____

8) The scale of a map of Texas is 4 inches: 32 miles. If you measure the distance from Dallas to Martin County as 38.4 inches, approximately how far is Martin County from Dallas? _____

# ACT Aspire Subject Test Mathematics Grade 8

## Ratio and Rates Word Problems

✍ **Find the answer for each word problem.**

1) Mason has 24 red cards and 36 green cards. What is the ratio of Mason's red cards to his green cards? _____

2) In a party, 45 soft drinks are required for every 54 guests. If there are 378 guests, how many soft drinks is required? _____

3) In Mason's class, 42 of the students are tall and 24 are short. In Michael's class 84 students are tall and 48 students are short. Which class has a higher ratio of tall to short students? _____

4) The price of 5 apples at the Quick Market is $4.6. The price of 7 of the same apples at Walmart is $5.95. Which place is the better buy? _____

5) The bakers at a Bakery can make 90 bagels in 3 hours. How many bagels can they bake in 24 hours? What is that rate per hour? _____

6) You can buy 5 cans of green beans at a supermarket for $5.75. How much does it cost to buy 45 cans of green beans? _____

7) The ratio of boys to girls in a class is 4: 7. If there are 32 boys in the class, how many girls are in that class? _____

8) The ratio of red marbles to blue marbles in a bag is 3: 7. If there are 50 marbles in the bag, how many of the marbles are red? _____

ACT Aspire Subject Test Mathematics Grade 8

## Percentage Calculations

✎ Calculate the given percent of each value.

1) 3% of 60 = ___
2) 20% of 32 = ___
3) 4% of 72 = ___
4) 16% of 32 = ___
5) 25% of 124 = ___
6) 35% of 56 = ___

7) 15% of 20 = ___
8) 14% of 150 = ___
9) 80% of 50 = ___
10) 12% of 115 = ___
11) 72% of 250 = ___
12) 52% of 500 = ___

13) 70% of 400 = ___
14) 27% of 145 = ___
15) 90% of 64 = ___
16) 60% of 55 = ___
17) 22% of 210 = ___
18) 8% of 235 = ___

✎ Calculate the percent of each given value.

19) ___% of 25 = 5
20) ___% of 40 = 20
21) ___% of 25 = 2
22) ___% of 50 = 16
23) ___% of 250 = 5

24) ___% of 40 = 32
25) ___% of 125 = 20
26) ___% of 700 = 49
27) ___% of 350 = 49
28) ___% of 500 = 210

✎ Calculate each percent problem.

29) A Cinema has 250 seats. 60 seats were sold for the current movie. What percent of seats are empty? ___ %

30) There are 68 boys and 92 girls in a class. 75% of the students in the class take the bus to school. How many students do not take the bus to school? ___

WWW.MathNotion.Com

# ACT Aspire Subject Test Mathematics Grade 8

## Percent Problems

✎ **Calculate each problem.**

1) 9 is what percent of 45? ____%

2) 60 is what percent of 120? ____%

3) 10 is what percent of 200? ____%

4) 15 is what percent of 125? ____%

5) 10 is what percent of 400? ____%

6) 66 is what percent of 55? ____%

7) 40 is what percent of 160? ____%

8) 40 is what percent of 50? ____%

9) 120 is what percent of 800? ____%

10) 78 is what percent of 120? ___%

11) 36 is what percent of 144? ___%

12) 17 is what percent of 85? ___%

13) 90 is what percent of 900? ___%

14) 36 is what percent of 16? ___%

15) 63 is what percent of 14? ___%

16) 18 is what percent of 60? ___%

17) 126 is what percent of 200? ___%

18) 232 is what percent of 40? ___%

✎ **Calculate each percent word problem.**

19) There are 40 employees in a company. On a certain day, 25 were present. What percent showed up for work? ____%

20) A metal bar weighs 60 ounces. 25% of the bar is gold. How many ounces of gold are in the bar? _____

21) A crew is made up of 12 women; the rest are men. If 15% of the crew are women, how many people are in the crew? _____

22) There are 40 students in a class and 8 of them are girls. What percent are boys? ____%

23) The Royals softball team played 400 games and won 280 of them. What percent of the games did they lose? ____%

ACT Aspire Subject Test Mathematics Grade 8

## Discount, Tax and Tip

✎ **Find the selling price of each item.**

1) Original price of a computer: $420
   Tax: 8%  Selling price: $_____

2) Original price of a laptop: $280
   Tax: 4%  Selling price: $_____

3) Original price of a sofa: $820
   Tax: 5%  Selling price: $_____

4) Original price of a car: $15,800
   Tax: 3.6%   Selling price: $_____

5) Original price of a Table: $250
   Tax: 9%  Selling price: $_____

6) Original price of a house: $630,000
   Tax: 1.8%   Selling price: $_____

7) Original price of a tablet: $450
   Discount: 30%   Selling price: $____

8) Original price of a chair: $390
   Discount: 8%   Selling price: $____

9) Original price of a book: $75
   Discount: 42%   Selling price: $____

10) Original price of a cellphone: $820
    Discount: 23%   Selling price: $___

11) Food bill: $45
    Tip: 15%        Price: $_____

12) Food bill: $32
    Tipp: 20%       Price: $_____

13) Food bill: $90
    Tip: 35%        Price: $_____

14) Food bill: $42
    Tipp: 12%   Price: $_____

✎ **Find the answer for each word problem.**

15) Nicolas hired a moving company. The company charged $500 for its services, and Nicolas gives the movers a 40% tip. How much does Nicolas tip the movers? $_____

16) Mason has lunch at a restaurant and the cost of his meal is $90. Mason wants to leave a 25% tip. What is Mason's total bill including tip? $_____

17) The sales tax in Texas is 19.80% and an item costs $350. How much is the tax? $_____

18) The price of a table at Best Buy is $680. If the sales tax is 5%, what is the final price of the table including tax? $_____

WWW.MathNotion.Com

# ACT Aspire Subject Test Mathematics Grade 8

## Percent of Change

### Find each percent of change.

1) From 150 to 450. ___ %
2) From 50 ft to 250 ft. ___ %
3) From $60 to $360. ___ %
4) From 60 cm to 180 cm. ___ %
5) From 15 to 45. ___ %
6) From 80 to 16. ___ %
7) From 120 to 360. ___ %
8) From 900 to 450. ___ %
9) From 1,000 to 200. ___ %
10) From 144 to 36. ___ %

### Calculate each percent of change word problem.

11) Bob got a raise, and his hourly wage increased from $42 to $63. What is the percent increase? ___ %

12) The price of a pair of shoes increases from $50 to $61. What is the percent increase? ___ %

13) At a coffee shop, the price of a cup of coffee increased from $4.80 to $5.76. What is the percent increase in the cost of the coffee? ___ %

14) 51 cm are cut from 85 cm board. What is the percent decrease in length? ___ %

15) In a class, the number of students has been increased from 54 to 81. What is the percent increase? ___ %

16) The price of gasoline rises from $24.40 to $30.50 in one month. By what percent did the gas price rise? ___ %

17) A shirt was originally priced at $38. It went on sale for $24.70. What was the percent that the shirt was discounted? ___ %

WWW.MathNotion.Com

ACT Aspire Subject Test Mathematics Grade 8

## Simple Interest

**Determine the simple interest for these loans.**

1) $480 at 11% for 3 years. $ _____
2) $4,200 at 7% for 4 years. $ _____
3) $2,500 at 20% for 3 years. $ _____
4) $6,800 at 3.9% for 4 months. $ _____
5) $800 at 6% for 7 months. $ _____
6) $36,000 at 4.2% for 6 years. $ _____
7) $6,500 at 7% for 4 years. $ _____
8) $850 at 9.5% for 2 years. $ _____
9) $1,200 at 5.8% for 9 months. $ _____
10) $3,000 at 4.5% for 7 years. $ _____

**Calculate each simple interest word problem.**

11) A new car, valued at $22,000, depreciates at 8.5% per year. What is the value of the car one year after purchase? $_____

12) Sara puts $9,000 into an investment yielding 6% annual simple interest; she left the money in for three years. How much interest does Sara get at the end of those three years? $_____

13) A bank is offering 12% simple interest on a savings account. If you deposit $16,400, how much interest will you earn in two years? $_____

14) $720 interest is earned on a principal of $6,000 at a simple interest rate of 4% interest per year. For how many years was the principal invested? _____

15) In how many years will $2,200 yield an interest of $440 at 4% simple interest? _____

16) Jim invested $8,000 in a bond at a yearly rate of 4.5%. He earned $1,440 in interest. How long was the money invested? _____

# ACT Aspire Subject Test Mathematics Grade 8

## Answers of Worksheets

**Simplifying Ratios**

1) 3 : 4
2) 1 : 10
3) 4 : 7
4) 1 : 3
5) 1 : 10
6) 1 : 8
7) 2 : 8
8) 2 : 5
9) 1 : 6
10) 7 : 9
11) 2 : 3
12) 7 : 2
13) 10 : 1
14) 3 : 4
15) 1 : 8
16) 5 : 7
17) 7 : 9
18) 2 : 1
19) 1 : 3
20) 7 : 1
21) 1 : 2
22) 5 : 4
23) 1 : 2
24) 1 : 30
25) $\frac{1}{2}$
26) $\frac{3}{5}$
27) $\frac{3}{7}$
28) $\frac{1}{3}$
29) $\frac{1}{3}$
30) $\frac{3}{14}$
31) $\frac{1}{2}$
32) $\frac{1}{5}$
33) $\frac{5}{12}$
34) $\frac{2}{9}$
35) $\frac{11}{13}$
36) $\frac{4}{9}$
37) $\frac{1}{12}$
38) $\frac{3}{11}$
39) $\frac{1}{15}$
40) $\frac{1}{4}$
41) $\frac{4}{3}$
42) $\frac{1}{5}$
43) $\frac{22}{41}$
44) $\frac{1}{9}$
45) $\frac{3}{5}$

**Proportional Ratios**

1) 18
2) 50
3) 9
4) 14
5) 14
6) 56
7) 100
8) 25
9) 70
10) 72
11) 200
12) 77
13) Yes
14) Yes
15) Yes
16) No
17) No
18) No
19) Yes
20) Yes
21) No
22) No
23) Yes
24) Yes
25) 40
26) 256
27) 7
28) 42
29) 63
30) 65
31) 20
32) 104
33) 55
34) 64
35) 126
36) 126

**Similarity and ratios**

1) 15
2) 5
3) 15
4) 13
5) 150 feet
6) 12.5 meters
7) 50 feet
8) 307.2 miles

**Ratio and Rates Word Problems**

1) 2 : 3
2) 315

# ACT Aspire Subject Test Mathematics Grade 8

3) The ratio for both classes is 7 to 4.

4) Walmart is a better buy.

5) 720, the rate is 30 per hour.

6) $51.75

7) 56

8) 15

**Percentage Calculations**

| | | |
|---|---|---|
| 1) 1.8 | 11) 180 | 21) 8% |
| 2) 6.4 | 12) 260 | 22) 32% |
| 3) 2.88 | 13) 280 | 23) 2% |
| 4) 5.12 | 14) 39.15 | 24) 80% |
| 5) 31 | 15) 57.6 | 25) 16% |
| 6) 19.6 | 16) 33 | 26) 7% |
| 7) 3 | 17) 46.2 | 27) 14% |
| 8) 21 | 18) 18.8 | 28) 42% |
| 9) 40 | 19) 20% | 29) 76% |
| 10) 13.8 | 20) 50% | 30) 40 |

**Percent Problems**

| | | |
|---|---|---|
| 1) 20% | 9) 15% | 17) 63% |
| 2) 50% | 10) 65% | 18) 580% |
| 3) 5% | 11) 25% | 19) 62.5% |
| 4) 12% | 12) 20% | 20) 15 ounces |
| 5) 2.5% | 13) 10% | 21) 80 |
| 6) 120% | 14) 225% | 22) 80% |
| 7) 25% | 15) 450% | 23) 30% |
| 8) 80% | 16) 30% | |

**Discount, Tax and Tip**

| | | |
|---|---|---|
| 1) $453.60 | 7) $315.00 | 13) $121.50 |
| 2) $291.20 | 8) $358.80 | 14) $47.04 |
| 3) $861.00 | 9) $43.50 | 15) $200.00 |
| 4) $16,368.80 | 10) $631.40 | 16) $112.50 |
| 5) $272.50 | 11) $51.75 | 17) $69.30 |
| 6) $641,340 | 12) $38.40 | 18) $714.00 |

# ACT Aspire Subject Test Mathematics Grade 8

**Percent of Change**

| | | |
|---|---|---|
| 1) 200% | 7) 200% | 13) 20% |
| 2) 400% | 8) 50% | 14) 60% |
| 3) 500% | 9) 80% | 15) 50% |
| 4) 200% | 10) 75% | 16) 25% |
| 5) 200% | 11) 50% | 17) 35% |
| 6) 80% | 12) 22% | |

**Simple Interest**

| | | |
|---|---|---|
| 1) $158.40 | 7) $1,820.00 | 13) $3,936.00 |
| 2) $1,176.00 | 8) $161.50 | 14) 3 years |
| 3) $1,500.00 | 9) $52.20 | 15) 5 years |
| 4) $88.40 | 10) $945.00 | 16) 4 years |
| 5) $28.00 | 11) $20,130.00 | |
| 6) $9,072.00 | 12) $1,620.00 | |

WWW.MathNotion.Com

# Chapter 4:
# Exponents and Radicals Expressions

**Topics that you will practice in this chapter:**

- ✓ Multiplication Property of Exponents
- ✓ Zero and Negative Exponents
- ✓ Division Property of Exponents
- ✓ Powers of Products and Quotients
- ✓ Negative Exponents and Negative Bases
- ✓ Scientific Notation
- ✓ Square Roots
- ✓ Simplifying Radical Expressions
- ✓ Simplifying Radical Expressions Involving Fractions
- ✓ Multiplying Radical Expressions
- ✓ Adding and Subtracting Radical Expressions

*Mathematics is no more computation than typing is literature.*

*– John Allen Paulos*

**ACT Aspire Subject Test Mathematics Grade 8**

## Multiplication Property of Exponents

✏️ **Simplify and write the answer in exponential form.**

1) $4 \times 4^5 =$

2) $8^4 \times 8 =$

3) $7^3 \times 7^3 =$

4) $9^2 \times 9^2 =$

5) $2^2 \times 2^4 \times 2 =$

6) $5 \times 5^3 \times 5^3 =$

7) $4^3 \times 4^2 \times 4 \times 4 =$

8) $5x \times x =$

9) $x^3 \times x^3 =$

10) $x^7 \times x^2 =$

11) $x^4 \times x^3 \times x^2 =$

12) $10x \times 3x =$

13) $4x^3 \times 4x^3 =$

14) $7x^3 \times x =$

15) $3x^2 \times 4x^2 \times x^2 =$

16) $5x^4 \times x^4 =$

17) $2x^8 \times 2x =$

18) $6x \times x^5 =$

19) $4x^2 \times 6x^6 =$

20) $5yx^3 \times 4x =$

21) $7x^3 \times y^5 x^7 =$

22) $y^2 x^3 \times y^5 x^4 =$

23) $3x^5 \times 4x^3 y^4 =$

24) $4x^4 \times 9x^2 y^5 =$

25) $5x^3 y^4 \times 6x^8 y^2 =$

26) $8x^3 y^6 \times 4xy^3 =$

27) $2xy^5 \times 6x^3 y^3 =$

28) $4x^5 y^2 \times 4x^2 y^8 =$

29) $7x \times 3y^8 x^2 \times y^5 =$

30) $x^3 \times 2y^3 x^4 \times 2y =$

31) $3yx^4 \times 3y^4 x \times 3xy^3 =$

32) $6y^3 \times 2y^2 x^4 \times 10yx^5 =$

WWW.MathNotion.Com

# ACT Aspire Subject Test Mathematics Grade 8

## Zero and Negative Exponents

✏️ **Evaluate the following expressions.**

1) $1^{-5} =$

2) $4^{-1} =$

3) $0^{10} =$

4) $1^{15} =$

5) $5^{-2} =$

6) $3^{-3} =$

7) $9^{-1} =$

8) $10^{-2} =$

9) $12^{-2} =$

10) $2^{-5} =$

11) $3^{-4} =$

12) $2^{-4} =$

13) $6^{-3} =$

14) $10^{-3} =$

15) $30^{-1} =$

16) $15^{-2} =$

17) $4^{-3} =$

18) $2^{-7} =$

19) $5^{-3} =$

20) $4^{-4} =$

21) $3^{-5} =$

22) $10^{-4} =$

23) $2^{-10} =$

24) $8^{-3} =$

25) $20^{-2} =$

26) $14^{-2} =$

27) $9^{-3} =$

28) $100^{-2} =$

29) $5^{-4} =$

30) $4^{-6} =$

31) $\left(\frac{1}{4}\right)^{-3} =$

32) $\left(\frac{1}{6}\right)^{-2} =$

33) $\left(\frac{1}{7}\right)^{-2} =$

34) $\left(\frac{2}{3}\right)^{-3} =$

35) $\left(\frac{1}{13}\right)^{-2} =$

36) $\left(\frac{7}{12}\right)^{-2} =$

37) $\left(\frac{1}{6}\right)^{-3} =$

38) $\left(\frac{1}{300}\right)^{-2} =$

39) $\left(\frac{2}{9}\right)^{-2} =$

40) $\left(\frac{7}{5}\right)^{-1} =$

41) $\left(\frac{13}{23}\right)^{0} =$

42) $\left(\frac{1}{4}\right)^{-5} =$

WWW.MathNotion.Com

# ACT Aspire Subject Test Mathematics Grade 8

## Division Property of Exponents

✎ Simplify.

1) $\dfrac{5^6}{5^7} =$

2) $\dfrac{8^8}{8^6} =$

3) $\dfrac{4^5}{4} =$

4) $\dfrac{3}{3^5} =$

5) $\dfrac{x}{x^6} =$

6) $\dfrac{3 \times 3^2}{3^2 \times 3^5} =$

7) $\dfrac{9^4}{9^2} =$

8) $\dfrac{10 \times 10^9}{10^2 \times 10^7} =$

9) $\dfrac{7^5 \times 7^7}{7^4 \times 7^8} =$

10) $\dfrac{15x}{30x^6} =$

11) $\dfrac{3x^9}{4x^4} =$

12) $\dfrac{15x^8}{10x^9} =$

13) $\dfrac{42x^5}{6y^9} =$

14) $\dfrac{36y^8}{4x^4y^5} =$

15) $\dfrac{2x^7}{9x} =$

16) $\dfrac{49x^8y^6}{7x^9} =$

17) $\dfrac{48x^2}{24x^6y^{12}} =$

18) $\dfrac{30yx^5}{6yx^7} =$

19) $\dfrac{19x^7y}{38x^{12}y^4} =$

20) $\dfrac{9x^8}{63x^8} =$

21) $\dfrac{9x^{-9}}{4x^{-3}} =$

WWW.MathNotion.Com

ACT Aspire Subject Test Mathematics Grade 8

## Powers of Products and Quotients

✏️ **Simplify.**

1) $(4^3)^2 =$

2) $(2^3)^4 =$

3) $(2 \times 2^3)^2 =$

4) $(5 \times 5^5)^6 =$

5) $(19^4 \times 19^2)^3 =$

6) $(2^3 \times 2^4)^4 =$

7) $(5 \times 5^2)^2 =$

8) $(4^4)^4 =$

9) $(8x^5)^2 =$

10) $(3x^2y^4)^4 =$

11) $(7x^5y^2)^2 =$

12) $(5x^4y^4)^3 =$

13) $(2x^3y^3)^5 =$

14) $(10x^3y^4)^3 =$

15) $(13y^3y)^2 =$

16) $(5x^6x^4)^2 =$

17) $(6x^7y^6)^3 =$

18) $(12x^5x^7)^2 =$

19) $(2x^4 \times 2x)^4 =$

20) $(2x^4y^3)^5 =$

21) $(15x^7y^2)^2 =$

22) $(8x^3y^5)^3 =$

23) $(3x \times 2y^2)^4 =$

24) $\left(\dfrac{4x}{x^5}\right)^2 =$

25) $\left(\dfrac{x^4y^5}{x^3y^5}\right)^9 =$

26) $\left(\dfrac{36xy}{6x^5}\right)^3 =$

27) $\left(\dfrac{x^7}{x^8y^2}\right)^6 =$

28) $\left(\dfrac{xy^4}{x^3y^6}\right)^{-3} =$

29) $\left(\dfrac{5xy^8}{x^3}\right)^2 =$

30) $\left(\dfrac{xy^6}{2xy^3}\right)^{-4} =$

WWW.MathNotion.Com

# ACT Aspire Subject Test Mathematics Grade 8

## Negative Exponents and Negative Bases

✏️ **Simplify.**

1) $-9^{-1} =$

2) $-9^{-2} =$

3) $-2^{-5} =$

4) $-x^{-7} =$

5) $11x^{-1} =$

6) $-8x^{-3} =$

7) $-12x^{-5} =$

8) $-9x^{-8}y^{-6} =$

9) $32x^{-5}y^{-1} =$

10) $10a^{-9}b^{-3} =$

11) $-17x^4y^{-6} =$

12) $-\dfrac{25}{x^{-5}} =$

13) $-\dfrac{13x}{a^{-7}} =$

14) $\left(-\dfrac{1}{3}\right)^{-4} =$

15) $\left(-\dfrac{3}{4}\right)^{-2} =$

16) $-\dfrac{14}{a^{-6}b^{-3}} =$

17) $-\dfrac{7x}{x^{-8}} =$

18) $-\dfrac{a^{-9}}{b^{-5}} =$

19) $-\dfrac{11}{x^{-5}} =$

20) $\dfrac{8b}{-16c^{-6}} =$

21) $\dfrac{12ab}{a^{-4}b^{-3}} =$

22) $-\dfrac{8n^{-4}}{32p^{-7}} =$

23) $\dfrac{16ab^{-6}}{-6c^{-5}} =$

24) $\left(\dfrac{10a}{5c}\right)^{-4} =$

25) $\left(-\dfrac{12x}{4yz}\right)^{-3} =$

26) $\dfrac{8ab^{-7}}{-5c^{-3}} =$

27) $\left(-\dfrac{x^4}{x^5}\right)^{-5} =$

28) $\left(-\dfrac{x^{-2}}{7x^3}\right)^{-2} =$

29) $\left(-\dfrac{x^{-4}}{x^2}\right)^{-6} =$

WWW.MathNotion.Com

ACT Aspire Subject Test Mathematics Grade 8

## Scientific Notation

✎ Write each number in scientific notation.

1) $0.223 =$

2) $0.09 =$

3) $4.5 =$

4) $900 =$

5) $2,000 =$

6) $0.006 =$

7) $33 =$

8) $9,400 =$

9) $1,470 =$

10) $52,000 =$

11) $8,000,000 =$

12) $0.00009 =$

13) $2,158,000 =$

14) $0.0039 =$

15) $0.000075 =$

16) $4,300,000 =$

17) $130,000 =$

18) $4,000,000,000 =$

19) $0.00009 =$

20) $0.0039 =$

✎ Write each number in standard notation.

21) $4 \times 10^{-1} =$

22) $1.2 \times 10^{-3} =$

23) $2.7 \times 10^{5} =$

24) $6 \times 10^{-4} =$

25) $3.6 \times 10^{-3} =$

26) $5.5 \times 10^{5} =$

27) $3.2 \times 10^{4} =$

28) $3.88 \times 10^{6} =$

29) $7 \times 10^{-6} =$

30) $4.2 \times 10^{-7} =$

WWW.MathNotion.Com

ACT Aspire Subject Test Mathematics Grade 8

## Square Roots

✎ Find the value each square root.

1) $\sqrt{16}$ = ___

2) $\sqrt{25}$ = ___

3) $\sqrt{1}$ = ___

4) $\sqrt{64}$ = ___

5) $\sqrt{0}$ = ___

6) $\sqrt{196}$ = ___

7) $\sqrt{4}$ = ___

8) $\sqrt{256}$ = ___

9) $\sqrt{36}$ = ___

10) $\sqrt{289}$ = ___

11) $\sqrt{169}$ = ___

12) $\sqrt{144}$ = ___

13) $\sqrt{100}$ = ___

14) $\sqrt{1,600}$ = ___

15) $\sqrt{2,500}$ = ___

16) $\sqrt{324}$ = ___

17) $\sqrt{529}$ = ___

18) $\sqrt{20}$ = ___

19) $\sqrt{625}$ = ___

20) $\sqrt{18}$ = ___

21) $\sqrt{50}$ = ___

22) $\sqrt{1,024}$ = ___

23) $\sqrt{160}$ = ___

24) $\sqrt{32}$ = ___

✎ Evaluate.

25) $\sqrt{4} \times \sqrt{25}$ = _____

26) $\sqrt{36} \times \sqrt{49}$ = _____

27) $\sqrt{6} \times \sqrt{6}$ = _____

28) $\sqrt{13} \times \sqrt{13}$ = _____

29) $2\sqrt{5} \times 3\sqrt{5}$ = _____

30) $\sqrt{12} \times \sqrt{3}$ = _____

31) $\sqrt{13} + \sqrt{13}$ = _____

32) $\sqrt{10} + 2\sqrt{10}$ = _____

33) $12\sqrt{7} - 10\sqrt{7}$ = _____

34) $4\sqrt{10} \times 2\sqrt{10}$ = _____

35) $5\sqrt{3} \times 8\sqrt{3}$ = _____

36) $6\sqrt{3} - \sqrt{12}$ = _____

WWW.MathNotion.Com

ACT Aspire Subject Test Mathematics Grade 8

## Simplifying Radical Expressions

✏️ **Simplify.**

1) $\sqrt{13x^2} =$

2) $\sqrt{75x^2} =$

3) $\sqrt[3]{27a} =$

4) $\sqrt{64x^5} =$

5) $\sqrt{216a} =$

6) $\sqrt[3]{63w^3} =$

7) $\sqrt{192x} =$

8) $\sqrt{125v} =$

9) $\sqrt[3]{128x^2} =$

10) $\sqrt{100x^9} =$

11) $\sqrt{16x^4} =$

12) $\sqrt[3]{500a^5} =$

13) $\sqrt{242} =$

14) $\sqrt{392p^3} =$

15) $\sqrt{8m^6} =$

16) $\sqrt{198x^3y^3} =$

17) $\sqrt{121x^5y^5} =$

18) $\sqrt{16a^6b^3} =$

19) $\sqrt{90x^5y^7} =$

20) $\sqrt[3]{64y^2x^6} =$

21) $10\sqrt{16x^4} =$

22) $6\sqrt{81x^2} =$

23) $\sqrt[3]{56x^2y^6} =$

24) $\sqrt[3]{1{,}000x^5y^7} =$

25) $8\sqrt{50a} =$

26) $\sqrt[4]{625x^8y} =$

27) $\sqrt{24x^4y^5r^3} =$

28) $5\sqrt{36x^4y^5z^8} =$

29) $3\sqrt[3]{343x^9y^7} =$

30) $5\sqrt{81a^5b^2c^9} =$

31) $\sqrt[4]{625x^8y^{16}} =$

WWW.MathNotion.Com

# ACT Aspire Subject Test Mathematics Grade 8

## Multiplying Radical Expressions

✎ **Simplify.**

1) $\sqrt{5} \times \sqrt{5} =$

2) $\sqrt{5} \times \sqrt{10} =$

3) $\sqrt{3} \times \sqrt{12} =$

4) $\sqrt{49} \times \sqrt{47} =$

5) $\sqrt{7} \times -2\sqrt{28} =$

6) $3\sqrt{15} \times \sqrt{5} =$

7) $4\sqrt{72} \times \sqrt{2} =$

8) $\sqrt{5} \times -\sqrt{49} =$

9) $\sqrt{55} \times \sqrt{11} =$

10) $7\sqrt{42} \times 2\sqrt{216} =$

11) $\sqrt{45}(5 + \sqrt{5}) =$

12) $\sqrt{13x^2} \times \sqrt{13x^3} =$

13) $-2\sqrt{27} \times \sqrt{3} =$

14) $2\sqrt{13x^4} \times \sqrt{13x^4} =$

15) $\sqrt{14x^3} \times \sqrt{7x^2} =$

16) $-8\sqrt{5x} \times \sqrt{7x^5} =$

17) $-2\sqrt{16x^5} \times 4\sqrt{8x^3} =$

18) $-4\sqrt{32}(8 + \sqrt{32}) =$

19) $\sqrt{32x}(10 - \sqrt{2x}) =$

20) $\sqrt{2x}(8\sqrt{x^5} + \sqrt{8}) =$

21) $\sqrt{20r}(5 + \sqrt{5}) =$

22) $-4\sqrt{7x} \times 3\sqrt{14x^5} =$

23) $-2\sqrt{12x} \times 3\sqrt{2x}$

24) $-\sqrt{7v^3}(-3\sqrt{42v}) =$

25) $(\sqrt{11} - 5)(\sqrt{11} + 5) =$

26) $(-3\sqrt{5} + 3)(\sqrt{5} - 4) =$

27) $(4 - 6\sqrt{3})(-6 + \sqrt{3}) =$

28) $(8 - 3\sqrt{5})(7 - \sqrt{5}) =$

29) $(-1 - \sqrt{3x})(4 + \sqrt{3x}) =$

30) $(-5 + 2\sqrt{7r})(-5 + \sqrt{7r}) =$

31) $(-\sqrt{7n} + 1)(-\sqrt{7} - 5) =$

32) $(-3 + \sqrt{3})(5 - 2\sqrt{3x}) =$

WWW.MathNotion.Com

# ACT Aspire Subject Test Mathematics Grade 8

## Simplifying Radical Expressions Involving Fractions

✍ **Simplify.**

1) $\dfrac{\sqrt{5}}{\sqrt{3}} =$

2) $\dfrac{\sqrt{18}}{\sqrt{45}} =$

3) $\dfrac{\sqrt{10}}{5\sqrt{2}} =$

4) $\dfrac{13}{\sqrt{3}} =$

5) $\dfrac{12\sqrt{5r}}{\sqrt{m^5}} =$

6) $\dfrac{11\sqrt{2}}{\sqrt{k}} =$

7) $\dfrac{6\sqrt{20x^3}}{\sqrt{16x}} =$

8) $\dfrac{\sqrt{14x^3 y^4}}{\sqrt{7x^4 y^3}} =$

9) $\dfrac{1}{1-\sqrt{5}} =$

10) $\dfrac{1-8\sqrt{a}}{\sqrt{11a}} =$

11) $\dfrac{\sqrt{a}}{\sqrt{a}+\sqrt{b}} =$

12) $\dfrac{1-\sqrt{5}}{2-\sqrt{6}} =$

13) $\dfrac{4+\sqrt{7}}{3-\sqrt{8}} =$

14) $\dfrac{5}{-3-3\sqrt{3}} =$

15) $\dfrac{7}{2-\sqrt{5}} =$

16) $\dfrac{\sqrt{7}-\sqrt{3}}{\sqrt{3}-\sqrt{7}} =$

17) $\dfrac{\sqrt{5}+\sqrt{7}}{\sqrt{7}-\sqrt{5}} =$

18) $\dfrac{2\sqrt{2}-\sqrt{3}}{3\sqrt{2}+\sqrt{5}} =$

19) $\dfrac{\sqrt{11}+5\sqrt{3}}{4-\sqrt{11}} =$

20) $\dfrac{\sqrt{5}+\sqrt{3}}{2-\sqrt{3}} =$

21) $\dfrac{\sqrt{32a^7 b^4}}{\sqrt{2ab^3}} =$

22) $\dfrac{10\sqrt{21x^5}}{5\sqrt{x^3}} =$

# ACT Aspire Subject Test Mathematics Grade 8

## Adding and Subtracting Radical Expressions

✏️ **Simplify.**

1) $\sqrt{2} + \sqrt{8} =$

2) $3\sqrt{50} + 4\sqrt{2} =$

3) $2\sqrt{12} - 4\sqrt{3} =$

4) $5\sqrt{32} - 5\sqrt{2} =$

5) $3\sqrt{75} - 5\sqrt{3} =$

6) $-\sqrt{72} - 4\sqrt{2} =$

7) $-7\sqrt{16} - 4\sqrt{25} =$

8) $8\sqrt{24} + 2\sqrt{6} =$

9) $10\sqrt{49} - 7\sqrt{100} =$

10) $-7\sqrt{5} + 9\sqrt{45} =$

11) $-15\sqrt{12} + 14\sqrt{48} =$

12) $20\sqrt{4} - 2\sqrt{25} =$

13) $-2\sqrt{20} + 7\sqrt{5} =$

14) $8\sqrt{7} - 2\sqrt{63} =$

15) $5\sqrt{44} + 3\sqrt{11} =$

16) $3\sqrt{27} - 5\sqrt{48} =$

17) $\sqrt{144} - \sqrt{81} =$

18) $3\sqrt{20} - 6\sqrt{5} =$

19) $-2\sqrt{7} + 8\sqrt{28} =$

20) $3\sqrt{75} - 2\sqrt{3} =$

21) $5\sqrt{27} - 3\sqrt{3} =$

22) $-7\sqrt{30} + 6\sqrt{120} =$

23) $-7\sqrt{24} - 2\sqrt{6} =$

24) $-\sqrt{32x} + 4\sqrt{2x} =$

25) $\sqrt{7y^2} + y\sqrt{112} =$

26) $\sqrt{45mn^2} + 2n\sqrt{5m} =$

27) $-4\sqrt{12a} - 4\sqrt{3a} =$

28) $-5\sqrt{15ab} - 2\sqrt{60ab} =$

29) $\sqrt{45x^2y} + x\sqrt{20y} =$

30) $2\sqrt{7a} + 4\sqrt{63a} =$

# ACT Aspire Subject Test Mathematics Grade 8

# Answers of Worksheets

**Multiplication Property of Exponents**

1) $4^6$
2) $8^5$
3) $7^6$
4) $9^4$
5) $2^7$
6) $5^7$
7) $4^7$
8) $5x^2$
9) $x^6$
10) $x^9$
11) $x^9$
12) $30x^2$
13) $16x^6$
14) $7x^4$
15) $12x^6$
16) $5x^8$
17) $4x^9$
18) $6x^6$
19) $24x^8$
20) $20x^4y$
21) $7x^{10}y^5$
22) $x^7y^7$
23) $12x^8y^4$
24) $36x^6y^5$
25) $30x^{11}y^6$
26) $32x^4y^9$
27) $12x^4y^8$
28) $16x^7y^{10}$
29) $21x^3y^{13}$
30) $4x^7y^4$
31) $27x^6y^8$
32) $120x^9y^6$

**Zero and Negative Exponents**

1) $1$
2) $\frac{1}{4}$
3) $0$
4) $1$
5) $\frac{1}{25}$
6) $\frac{1}{27}$
7) $\frac{1}{9}$
8) $\frac{1}{100}$
9) $\frac{1}{144}$
10) $\frac{1}{32}$
11) $\frac{1}{81}$
12) $\frac{1}{16}$
13) $\frac{1}{216}$
14) $\frac{1}{1,000}$
15) $\frac{1}{30}$
16) $\frac{1}{225}$
17) $\frac{1}{64}$
18) $\frac{1}{128}$
19) $\frac{1}{125}$
20) $\frac{1}{256}$
21) $\frac{1}{243}$
22) $\frac{1}{10,000}$
23) $\frac{1}{1,024}$
24) $\frac{1}{512}$
25) $\frac{1}{400}$
26) $\frac{1}{196}$
27) $\frac{1}{729}$
28) $\frac{1}{10,000}$
29) $\frac{1}{625}$
30) $\frac{1}{4,096}$
31) $64$
32) $36$
33) $49$
34) $\frac{27}{8}$
35) $169$
36) $\frac{144}{49}$
37) $216$
38) $90,000$
39) $\frac{81}{4}$
40) $\frac{5}{7}$
41) $1$
42) $1,024$

**Division Property of Exponents**

1) $\frac{1}{5}$
2) $8^2$
3) $4^4$
4) $\frac{1}{3^4}$
5) $\frac{1}{x^5}$
6) $\frac{1}{3^4}$
7) $9^2$
8) $10$
9) $1$
10) $\frac{1}{2x^5}$
11) $\frac{3x^5}{4}$
12) $\frac{3}{2x}$
13) $\frac{7x^5}{y^9}$
14) $\frac{9y^3}{x^4}$
15) $\frac{2x^6}{9}$

# ACT Aspire Subject Test Mathematics Grade 8

16) $\frac{7y^6}{x}$     18) $\frac{5}{x^2}$     19) $\frac{1}{2x^5y^3}$     21) $\frac{9}{4x^6}$

17) $\frac{2}{x^4y^{12}}$                    20) $\frac{1}{7}$

## Powers of Products and Quotients

1) $4^6$
2) $2^{12}$
3) $2^8$
4) $5^{36}$
5) $19^{18}$
6) $2^{28}$
7) $5^6$
8) $4^{16}$
9) $64x^{10}$
10) $81x^8y^{16}$
11) $49x^{10}y^4$
12) $125x^{12}y^{12}$
13) $32x^{15}y^{15}$
14) $1{,}000x^9y^{12}$
15) $169y^8$
16) $25x^{20}$
17) $216x^{21}y^{18}$
18) $144x^{24}$
19) $256x^{20}$
20) $32x^{20}y^{15}$
21) $225x^{14}y^4$
22) $512x^9y^{15}$
23) $1{,}296x^4y^8$
24) $\frac{16}{x^8}$
25) $x^9$
26) $\frac{216y^3}{x^{12}}$
27) $\frac{1}{x^6y^{12}}$
28) $x^6y^6$
29) $\frac{25y^{16}}{x^4}$
30) $\frac{16}{y^{12}}$

## Negative Exponents and Negative Bases

1) $-\frac{1}{9}$
2) $-\frac{1}{81}$
3) $-\frac{1}{32}$
4) $-\frac{1}{x^7}$
5) $\frac{11}{x}$
6) $-\frac{8}{x^3}$
7) $-\frac{12}{x^5}$
8) $-\frac{9}{x^8y^6}$
9) $\frac{32}{x^5y}$
10) $\frac{10}{a^9b^3}$
11) $-\frac{17x^4}{y^6}$
12) $-25x^5$
13) $-13xa^7$
14) $81$
15) $\frac{16}{9}$
16) $-14a^6b^3$
17) $-7x^9$
18) $-\frac{b^5}{a^9}$
19) $-11x^5$
20) $-\frac{bc^6}{2}$
21) $12a^5b^4$
22) $-\frac{p^7}{4n^4}$
23) $-\frac{8ac^5}{3b^6}$
24) $\frac{c^4}{16a^4}$
25) $\frac{y^3z^3}{27x^3}$
26) $-\frac{8ac^3}{5b^7}$
27) $-x^5$
28) $49x^{10}$
29) $x^{36}$

## Scientific Notation

WWW.MathNotion.Com

# ACT Aspire Subject Test Mathematics Grade 8

1) $2.23 \times 10^{-1}$
2) $9 \times 10^{-2}$
3) $4.5 \times 10^{0}$
4) $9 \times 10^{2}$
5) $2 \times 10^{3}$
6) $6 \times 10^{-3}$
7) $3.3 \times 10^{1}$
8) $9.4 \times 10^{3}$
9) $1.47 \times 10^{3}$
10) $5.2 \times 10^{4}$
11) $8 \times 10^{6}$
12) $9 \times 10^{-5}$
13) $2.158 \times 10^{6}$
14) $3.9 \times 10^{-3}$
15) $7.5 \times 10^{-5}$
16) $4.3 \times 10^{6}$
17) $1.3 \times 10^{5}$
18) $4 \times 10^{9}$
19) $9 \times 10^{-5}$
20) $3.9 \times 10^{-3}$
21) 0.4
22) 0.0012
23) 270,000
24) 0.0006
25) 0.0036
26) 550,000
27) 32,000
28) 3,880,000
29) 0.000007
30) 0.00000042

**Square Roots**

1) 4
2) 5
3) 1
4) 8
5) 0
6) 14
7) 2
8) 16
9) 6
10) 17
11) 13
12) 12
13) 10
14) 40
15) 50
16) 18
17) 23
18) $2\sqrt{5}$
19) 25
20) $3\sqrt{2}$
21) $5\sqrt{2}$
22) 32
23) $4\sqrt{10}$
24) $4\sqrt{2}$
25) 10
26) 42
27) 6
28) 13
29) 30
30) 6
31) $2\sqrt{13}$
32) $3\sqrt{10}$
33) $2\sqrt{7}$
34) 80
35) 120
36) $4\sqrt{3}$

**Simplifying radical expressions**

1) $x\sqrt{13}$
2) $5x\sqrt{3}$
3) $3\sqrt[3]{a}$
4) $8x^2\sqrt{x}$
5) $6\sqrt{6a}$
6) $w\sqrt[3]{63}$
7) $8\sqrt{3x}$
8) $5\sqrt{5v}$
9) $4\sqrt[3]{2x^2}$
10) $10x^4\sqrt{x}$
11) $4x^2$
12) $5a\sqrt[3]{4a^2}$
13) $11\sqrt{2}$
14) $14p\sqrt{2p}$
15) $2m^3\sqrt{2}$
16) $3x.y\sqrt{22xy}$
17) $11x^2y^2\sqrt{xy}$
18) $4a^3b\sqrt{b}$
19) $3x^2y^3\sqrt{10xy}$
20) $4x^2\sqrt[3]{y^2}$
21) $40x^2$
22) $54x$
23) $2y^2\sqrt[3]{7x^2}$
24) $10xy^2\sqrt[3]{x^2y}$

# ACT Aspire Subject Test Mathematics Grade 8

25) $40\sqrt{2a}$

26) $5x^2\sqrt[4]{y}$

27) $2x^2y^2r\sqrt{6yr}$

28) $30x^2y^2z^4\sqrt{y}$

29) $21x^3y^2\sqrt[3]{y}$

30) $45a^2bc^4\sqrt{ac}$

31) $5x^2y^4$

**Multiplying radical expressions**

1) 5

2) $5\sqrt{2}$

3) 6

4) $7\sqrt{47}$

5) $-28$

6) $15\sqrt{3}$

7) 48

8) $-5\sqrt{7}$

9) $11\sqrt{5}$

10) $504\sqrt{7}$

11) $15\sqrt{5} + 15$

12) $13x^2\sqrt{x}$

13) $-18$

14) $26x^4$

15) $7x^2\sqrt{2x}$

16) $-8x^3\sqrt{35}$

17) $-64x^4\sqrt{2}$

18) $-128\sqrt{2} - 128$

19) $40\sqrt{2x} - 8x$

20) $8x^3\sqrt{2} + 4\sqrt{x}$

21) $10\sqrt{5r} + 10\sqrt{r}$

22) $-84x^3\sqrt{2}$

23) $-12\sqrt{6}x$

24) $21v^2\sqrt{6}$

25) $-14$

26) $15\sqrt{5} - 27$

27) $40\sqrt{3} - 42$

28) $71 - 29\sqrt{5}$

29) $-3x - 5\sqrt{3x} - 4$

30) $14r - 15\sqrt{7r} + 25$

31) $7\sqrt{n} + 5\sqrt{7n} - \sqrt{7} - 5$

32) $-15 + 6\sqrt{3x} + 5\sqrt{3} - 6\sqrt{x}$

**Simplifying radical expressions involving fractions**

1) $\frac{\sqrt{15}}{3}$

2) $\frac{9\sqrt{10}}{45} = \frac{\sqrt{10}}{5}$

3) $\frac{\sqrt{20}}{10} = \frac{\sqrt{5}}{5}$

4) $\frac{13\sqrt{3}}{3}$

5) $\frac{12\sqrt{5mr}}{m^3}$

6) $\frac{11\sqrt{2k}}{k}$

7) $3x\sqrt{5}$

8) $\frac{\sqrt{2x}}{xy}$

9) $\frac{-1-\sqrt{5}}{4}$

10) $\frac{\sqrt{11a} - 8a\sqrt{11}}{11a}$

# ACT Aspire Subject Test Mathematics Grade 8

11) $\dfrac{a-\sqrt{ab}}{a-b}$

12) $\dfrac{\sqrt{30}+2\sqrt{5}-\sqrt{6}-2}{2}$

13) $12 + 8\sqrt{2} + 3\sqrt{7} + 2\sqrt{14}$

14) $-\dfrac{5(\sqrt{3}-1)}{6}$

15) $-14 - 7\sqrt{5}$

16) $-1$

17) $6 + \sqrt{35}$

18) $\dfrac{12 - 2\sqrt{10} - 3\sqrt{6} + \sqrt{15}}{13}$

19) $\dfrac{4\sqrt{11}+11+20\sqrt{3}+5\sqrt{33}}{5}$

20) $2\sqrt{5} + 3 + \sqrt{15} + 2\sqrt{3}$

21) $4a^3\sqrt{b}$

22) $2x\sqrt{21}$

**Adding and subtracting radical expressions**

1) $3\sqrt{2}$
2) $19\sqrt{2}$
3) $0$
4) $15\sqrt{2}$
5) $10\sqrt{3}$
6) $-10\sqrt{2}$
7) $-48$
8) $18\sqrt{6}$
9) $0$
10) $20\sqrt{5}$

11) $26\sqrt{3}$
12) $30$
13) $3\sqrt{5}$
14) $2\sqrt{7}$
15) $13\sqrt{11}$
16) $-11\sqrt{3}$
17) $3$
18) $0$
19) $14\sqrt{7}$
20) $13\sqrt{3}$

21) $12\sqrt{3}$
22) $5\sqrt{30}$
23) $-16\sqrt{6}$
24) $0$
25) $5y\sqrt{7}$
26) $5n\sqrt{5m}$
27) $-12\sqrt{3a}$
28) $-9\sqrt{15ab}$
29) $5x\sqrt{5y}$
30) $14\sqrt{7a}$

**ACT Aspire Subject Test Mathematics Grade 8**

# Chapter 5 :
# Algebraic Expressions

## Topics that you will practice in this chapter:

- ✓ Simplifying Variable Expressions
- ✓ Simplifying Polynomial Expressions
- ✓ Translate Phrases into an Algebraic Statement
- ✓ The Distributive Property
- ✓ Evaluating One Variable Expressions
- ✓ Evaluating Two Variables Expressions
- ✓ Combining like Terms

*Mathematics is, as it were, a sensuous logic, and relates to philosophy as do the arts, music, and plastic art to poetry.* — *K. Shegel*

ACT Aspire Subject Test Mathematics Grade 8

## Simplifying Variable Expressions

✏️ **Simplify each expression.**

1) $3(x + 5) =$

2) $(-4)(7x - 5) =$

3) $11x + 5 - 6x =$

4) $-4 - 2x^2 - 6x^2 =$

5) $7 + 13x^2 + 3 =$

6) $3x^2 + 7x + 15x^2 =$

7) $3x^2 - 12x^2 + 4x =$

8) $4x^2 - 8x - 2x =$

9) $6x + 7(3 - 4x) =$

10) $8x + 4(15x - 3) =$

11) $6(-3x - 9) - 17 =$

12) $-11x^2 - (-5x) =$

13) $2x + 7 + 5 - 8x =$

14) $7 + 6x - 11 - 5x =$

15) $27x + 8 - 13 - 5x =$

16) $(-11)(-5x + 2) - 41x =$

17) $19x - 4(4 - 2x) =$

18) $16x + 3(3x + 6) + 10 =$

19) $5(-2x - 4) - 13x =$

20) $16x - 3x(x + 10) =$

21) $17x + 5x(2 - 4x) =$

22) $5x(-4x - 7) + 20x =$

23) $25x - 19 + 4x^2 =$

24) $6x(x - 11) + 25 =$

25) $4x - 5 + 15x + 3x^2 =$

26) $-7x^2 - 11x - 9x =$

27) $10x - 9x^2 - 3x^2 - 7 =$

28) $13 + 3x^2 - 9x^2 - 21x =$

29) $22x + 10x^2 - 15x + 17 =$

30) $4x^2 + 25x + 21x^2 =$

31) $29 - 12x^2 - 23x - 4x^2 =$

32) $22x - 19x - 9x^2 + 30 =$

ACT Aspire Subject Test Mathematics Grade 8

## Simplifying Polynomial Expressions

✏️ **Simplify each polynomial.**

1) $(2x^3 + 8x^2) - (11x + 3x^2) = $ _____

2) $(2x^5 + 7x^3) - (5x^3 + 11x^2) = $ _____

3) $(41x^4 + 5x^2) - (4x^2 + 20x^4) = $ _____

4) $13x - 8x^2 + 4(4x^2 + 3x^3) = $ _____

5) $(4x^3 - 22) + 5(3x^2 - 6x^3) = $ _____

6) $(4x^3 - 3x) - 5(2x^3 + x^4) = $ _____

7) $5(5x - 2x^3) - 2(8x^3 + 5x^2) = $ _____

8) $(3x^2 - 10x) - (5x^3 + 14x^2) = $ _____

9) $5x^3 - (3x^4 + 5x) + 2x^2 = $ _____

10) $11x^4 - (3x^2 + 5x) + 7x = $ _____

11) $(6x^2 - 3x^4) - (10x^4 + 3x^2) = $ _____

12) $2x^2 - 7x^3 + 19x^4 - 22x^3 = $ _____

13) $10x^2 - x^4 + 4x^4 - 32x^3 = $ _____

14) $-5x^2 + 17x^3 - 8x^2 - 6x = $ _____

15) $x^4 - 11x^5 - 30x^4 + 5x^2 = $ _____

16) $21x^3 + 13x - 5x^2 - 11x^3 = $ _____

WWW.MathNotion.Com

ACT Aspire Subject Test Mathematics Grade 8

## Translate Phrases into an Algebraic Statement

✏️ **Write an algebraic expression for each phrase.**

1) 9 multiplied by $x$. _____

2) Subtract 11 from $y$. _____

3) 19 divided by $x$. _____

4) 38 decreased by $y$. _____

5) Add $y$ to 40. _____

6) The square of 6. _____

7) $x$ raised to the fifth power. _____

8) The sum of six and a number. _____

9) The difference between fifty-seven and $y$. _____

10) The quotient of nine and a number. _____

11) The quotient of the square of $x$ and 25. _____

12) The difference between $x$ and 6 is 19. _____

13) 10 times $a$ reduced by the square of $b$. _____

14) Subtract the product of $a$ and $b$ from 41. _____

# ACT Aspire Subject Test Mathematics Grade 8

## The Distributive Property

✏️ Use the distributive property to simply each expression.

1) $4(1 + 2x) =$

2) $2(4 + 7x) =$

3) $3(4x - 4) =$

4) $(2x - 5)(-6) =$

5) $(-3)(x + 6) =$

6) $(4 + 3x)2 =$

7) $(-5)(8 - 3x) =$

8) $-(-5 - 7x) =$

9) $(-6x + 3)(-3) =$

10) $(-4)(x - 7) =$

11) $-(5 - 3x) =$

12) $3(9 + 4x) =$

13) $6(4 + 3x) =$

14) $(-5x + 3)2 =$

15) $(5 - 8x)(-3) =$

16) $(-12)(3x + 3) =$

17) $(5 - 3x)6 =$

18) $4(2 + 6x) =$

19) $8(7x - 3) =$

20) $(-2x + 3)4 =$

21) $(7 - 5x)(-9) =$

22) $(-10)(x - 8) =$

23) $(11 - 4x)3 =$

24) $(-6)(10x - 4) =$

25) $(3 - 9x)(-7) =$

26) $(-9)(x + 9) =$

27) $(-3 + 5x)(-7) =$

28) $(-5)(8 - 10x) =$

29) $12(4x - 8) =$

30) $(-10x + 13)(-3) =$

31) $(-8)(3x - 2) + 4(x + 5) =$

32) $(-8)(x + 4) - (6 + 5x) =$

WWW.MathNotion.Com

ACT Aspire Subject Test Mathematics Grade 8

## Evaluating One Variable Expressions

✏ **Evaluate each expression using the value given.**

1) $8 - x, x = 5$

2) $x - 9, x = 5$

3) $5x + 4, x = 3$

4) $x - 13, x = -4$

5) $12 - x, x = 4$

6) $x + 2, x = 6$

7) $4x + 8, x = 3$

8) $x + (-7), x = -8$

9) $4x + 5, x = 2$

10) $3x + 9, x = -2$

11) $15 + 3x - 7, x = 2$

12) $17 - 3x, x = 3$

13) $8x - 9, x = 4$

14) $5x + 4, x = -3$

15) $10x + 5, x = 3$

16) $14 - 4x, x = -6$

17) $3(5x + 3), x = 9$

18) $4(-3x - 6), x = 3$

19) $7x - 2x + 12, x = 4$

20) $(5x + 6) \div 2, x = 8$

21) $(x + 18) \div 10, x = 12$

22) $5x - 12 + 3x, x = -3$

23) $(6 - 4x)(-3), x = -4$

24) $9x^2 + 3x - 6, x = 2$

25) $x^2 - 10x, x = -5$

26) $3x(7 - 2x), x = 2$

27) $12x + 6 - 2x^2, x = -4$

28) $(-3)(4x - 8 + 3x), x = 3$

29) $(-6) + \frac{x}{4} + 3x, x = 16$

30) $(-6) + \frac{x}{5}, x = 35$

31) $\left(-\frac{45}{x}\right) - 7 + 2x, x = 9$

32) $\left(-\frac{21}{x}\right) - 12 + 4x, x = 7$

WWW.MathNotion.Com

ACT Aspire Subject Test Mathematics Grade 8

## Evaluating Two Variables Expressions

✎ Evaluate each expression using the values given.

1) $2x - 4y$,
   $x = 4, y = 1$

2) $3x + 5y$,
   $x = -2, y = 2$

3) $-7a + 4b$,
   $a = 2, b = 4$

4) $3x + 5 - y$,
   $x = 5, y = 6$

5) $3z + 12 - 2k$,
   $z = 5, k = 6$

6) $6(-x - 3y)$,
   $x = 5, y = -2$

7) $5a + 3b$,
   $a = 3, b = 4$

8) $7x \div 3y$,
   $x = 3, y = 7$

9) $2x + 15 + 5y$,
   $x = -3, y = 1$

10) $5a - (18 - b)$,
    $a = 2, b = 8$

11) $2z + 20 + 5k$,
    $z = -6, k = 5$

12) $xy + 10 + 4x$,
    $x = 3, y = 5$

13) $2x + 4y - 8 + 5$,
    $x = 5, y = 2$

14) $\left(-\frac{24}{x}\right) + 3 + 2y$,
    $x = 4, y = 6$

15) $(-3)(-3a - 3b)$,
    $a = 4, b = 5$

16) $12 + 4x - 7 - y$,
    $x = 3, y = 5$

17) $11x + 5 - 8y + 6$,
    $x = 5, y = 2$

18) $10 + 2(-4x - 5y)$,
    $x = 5, y = 4$

19) $5x + 13 + 6y$,
    $x = 5, y = 6$

20) $10a - (7a + 3b) - 11$,
    $a = 3, b = 8$

WWW.MathNotion.Com

# ACT Aspire Subject Test Mathematics Grade 8

## Combining like Terms

✏️ **Simplify each expression.**

1) $11x + 3x + 6 =$

2) $8(2x - 6) =$

3) $18x - 7x + 11 =$

4) $(-4)(6x - 7) =$

5) $22x - 10x - 5 =$

6) $32x - 13 + 8x =$

7) $15 - (8x - 11) =$

8) $-24x + 17 - 11x =$

9) $12x - 8 - 6x + 9 =$

10) $21x + 5 - 36 + 12x =$

11) $28x + 3x - 11 =$

12) $(-3x + 4)5 =$

13) $2 + 4x + 9x - 8 =$

14) $6(2x - 5x) - 4 =$

15) $4(5x + 11) + 3x =$

16) $x - 14 - 11x =$

17) $5(10 + 9x) - 8x =$

18) $42x + 17 - 23x =$

19) $(-7x) + 19 + 20x =$

20) $(-7x) - 33 + 29x =$

21) $4(5x + 3) - 19x =$

22) $5(6 - 2x) - 15x =$

23) $-24x + (11 - 18x) =$

24) $(-9) - (6)(7x + 3) =$

25) $(-1)(8x - 10) - 21x =$

26) $-36x + 14 + 27x - 5x =$

27) $3(-13x + 6) - 17x =$

28) $-5x - 42 + 32x =$

29) $37x - 19x + 15 - 9x =$

30) $3(5x + 7x) - 31 =$

31) $14 - 6x - 15 - 9x =$

32) $-2(-5x - 7x) + 27x =$

WWW.MathNotion.Com

**ACT Aspire Subject Test Mathematics Grade 8**

## Answers of Worksheets

**Simplifying Variable Expressions**

1) $3x + 15$
2) $-28x + 20$
3) $5x + 5$
4) $-8x^2 - 4$
5) $13x^2 + 10$
6) $18x^2 + 7x$
7) $-9x^2 + 4x$
8) $4x^2 - 10x$
9) $-22x + 21$
10) $68x - 12$
11) $-18x - 71$
12) $-11x^2 + 5x$
13) $-6x + 12$
14) $x - 4$
15) $22x - 5$
16) $14x - 22$
17) $27x - 16$
18) $25x + 28$
19) $-23x - 20$
20) $-3x^2 - 14x$
21) $-20x^2 + 27x$
22) $-20x^2 - 15x$
23) $4x^2 + 25x - 19$
24) $6x^2 - 66x + 25$
25) $3x^2 + 19x - 5$
26) $-7x^2 - 20x$
27) $-12x^2 + 10x - 7$
28) $-6x^2 - 21x + 13$
29) $10x^2 + 7x + 17$
30) $25x^2 + 25x$
31) $-16x^2 - 23x + 29$
32) $-9x^2 + 3x + 30$

**Simplifying Polynomial Expressions**

1) $2x^3 + 5x^2 - 11x$
2) $2x^5 + 2x^3 - 11x^2$
3) $21x^4 + x^2$
4) $12x^3 + 8x^2 + 13x$
5) $-26x^3 + 15x^2 - 22$
6) $-5x^4 - 6x^3 - 3x$
7) $-26x^3 - 10x^2 + 25x$
8) $-5x^3 - 11x^2 - 10x$
9) $-3x^4 + 5x^3 + 2x^2 - 5x$
10) $11x^4 - 3x^2 + 2x$
11) $-13x^4 + 3x^2$
12) $19x^4 - 29x^3 + 2x^2$
13) $3x^4 - 32x^3 + 10x^2$
14) $17x^3 - 13x^2 - 6x$
15) $-11x^5 - 29x^4 + 5x^2$
16) $10x^3 - 5x^2 + 13x$

**Translate Phrases into an Algebraic Statement**

1) $9x$
2) $y - 11$
3) $\frac{19}{x}$
4) $38 - y$
5) $y + 40$
6) $6^2$
7) $x^5$
8) $6 + x$
9) $57 - y$
10) $\frac{9}{x}$
11) $\frac{x^2}{25}$
12) $x - 6 = 19$
13) $10a - b^2$
14) $41 - ab$

**The Distributive Property**

1) $8x + 4$
2) $14x + 8$
3) $12x - 12$
4) $-12x + 30$
5) $-3x - 18$
6) $6x + 8$
7) $15x - 40$
8) $7x + 5$
9) $18x - 9$
10) $-4x + 28$
11) $3x - 5$
12) $12x + 27$

# ACT Aspire Subject Test Mathematics Grade 8

13) $18x + 24$
14) $-10x + 6$
15) $24x - 15$
16) $-36x - 36$
17) $-18x + 30$

18) $24x + 8$
19) $56x - 24$
20) $-8x + 12$
21) $45x - 63$
22) $-10x + 80$

23) $-12x + 33$
24) $-60x + 24$
25) $63x - 21$
26) $-9x - 81$
27) $-35x + 21$

28) $50x - 40$
29) $48x - 96$
30) $30x - 39$
31) $-20x + 36$
32) $-13x - 38$

**Evaluating One Variables**

1) 3
2) $-4$
3) 19
4) $-17$
5) 8
6) 8
7) 20
8) $-15$

9) 13
10) 3
11) 14
12) 8
13) 23
14) $-11$
15) 35
16) 38

17) 144
18) $-60$
19) 32
20) 23
21) 3
22) $-36$
23) $-66$
24) 36

25) 75
26) 18
27) $-74$
28) $-39$
29) 46
30) 1
31) 6
32) 13

**Evaluating Two Variables**

1) 4
2) 4
3) 2
4) 14
5) 15

6) 6
7) 27
8) 1
9) 14
10) 0

11) 33
12) 37
13) 15
14) 9
15) 81

16) 12
17) 50
18) $-70$
19) 74
20) $-26$

**Combining like Terms**

1) $14x + 6$
2) $16x - 48$
3) $11x + 11$
4) $-24x + 28$
5) $12x - 5$
6) $40x - 13$
7) $-8x + 26$
8) $-35x + 17$

9) $6x + 1$
10) $33x - 31$
11) $31x - 11$
12) $-15x + 20$
13) $13x - 6$
14) $-18x - 4$
15) $23x + 44$
16) $-10x - 14$

17) $37x + 50$
18) $19x + 17$
19) $13x + 19$
20) $22x - 33$
21) $x + 12$
22) $-25x + 30$
23) $-42x + 11$
24) $-42x - 27$

25) $-29x + 10$
26) $-14x + 14$
27) $-56x + 18$
28) $27x - 42$
29) $9x + 15$
30) $36x - 31$
31) $-15x - 1$
32) $51x$

WWW.MathNotion.Com

ACT Aspire Subject Test Mathematics Grade 8

# Chapter 6
# Equations and Inequalities

**Topics that you will practice in this chapter:**

- ✓ One-Step Equations
- ✓ Multi-Step Equations
- ✓ Graphing Single-Variable Inequalities
- ✓ One-Step Inequalities
- ✓ Multi-Step Inequalities
- ✓ Systems of Equations
- ✓ Systems of Equations Word Problems

*"Life is a math equation. In order to gain the most, you have to know how to convert negatives into positives." – Anonymous*

ACT Aspire Subject Test Mathematics Grade 8

## One–Step Equations

✎ **Find the answer for each equation.**

1) $3x = 90, x =$ \_\_\_\_

2) $5x = 35, x =$ \_\_\_\_

3) $6x = 24, x =$ \_\_\_\_

4) $24x = 144, x =$ \_\_\_\_

5) $x + 15 = 20, x =$ \_\_\_\_

6) $x - 7 = 4, x =$ \_\_\_\_

7) $x - 9 = 2, x =$ \_\_\_\_

8) $x + 15 = 23, x =$ \_\_\_\_

9) $x - 4 = 13, x =$ \_\_\_\_

10) $12 = 16 + x, x =$ \_\_\_\_

11) $x - 10 = 2, x =$ \_\_\_\_

12) $5 - x = -11, x =$ \_\_\_\_

13) $28 = -6 + x, x =$ \_\_\_\_

14) $x - 20 = -35, x =$ \_\_\_\_

15) $x + 14 = -4, x =$ \_\_\_\_

16) $14 = 28 - x, x =$ \_\_\_\_

17) $7 + x = -7, x =$ \_\_\_\_

18) $x - 16 = 4, x =$ \_\_\_\_

19) $30 = x - 15, x =$ \_\_\_\_

20) $x - 5 = -18, x =$ \_\_\_\_

21) $x - 10 = 24, x =$ \_\_\_\_

22) $x - 20 = -25, x =$ \_\_\_\_

23) $x - 17 = 30, x =$ \_\_\_\_

24) $-70 = x - 28, x =$ \_\_\_\_

25) $x - 9 = 13, x =$ \_\_\_\_

26) $36 = 4x, x =$ \_\_\_\_

27) $x - 35 = 25, x =$ \_\_\_\_

28) $x - 25 = 10, x =$ \_\_\_\_

29) $70 - x = 16, x =$ \_\_\_\_

30) $x - 10 = 14, x =$ \_\_\_\_

31) $17 - x = -13, x =$ \_\_

32) $x - 9 = -30, x =$ \_\_\_\_

WWW.MathNotion.Com

ACT Aspire Subject Test Mathematics Grade 8

## Multi–Step Equations

✏ **Find the answer for each equation.**

1) $3x + 3 = 9$

2) $-x + 5 = 12$

3) $4x - 8 = 8$

4) $-(3 - x) = 5$

5) $4x - 8 = 16$

6) $12x - 15 = 9$

7) $2x - 18 = 2$

8) $4x + 8 = 16$

9) $24x + 27 = 75$

10) $-14(3 + x) = 14$

11) $-3(2 + x) = 6$

12) $12 = -(x - 7)$

13) $3(3 - x) = 30$

14) $-15 = -(3x + 6)$

15) $40(3 + x) = 40$

16) $5(x - 10) = 25$

17) $-18 = x + 8x$

18) $3x + 25 = -2x - 10$

19) $7(6 + 3x) = -63$

20) $18 - 3x = -4 - 5x$

21) $4 - 6x = 36 + 2x$

22) $15 + 15x = -5 + 5x$

23) $42 = (-6x) - 7 + 7$

24) $21 = 3x - 21 + 4x$

25) $-18 = -6x - 9 + 3x$

26) $5x - 15 = -29 + 6x$

27) $7x - 18 = 4x + 3$

28) $-7 - 4x = 5(4 - x)$

29) $x - 5 = -5(-3 - x)$

30) $13x - 68 = 15x - 102$

31) $-5x - 3 = -3(9 + 3x)$

32) $-2x - 15 = 6x + 17$

WWW.MathNotion.Com

# ACT Aspire Subject Test Mathematics Grade 8

## Graphing Single–Variable Inequalities

✎ **Draw a graph for each inequality.**

1) $x > -1$

2) $x \leq 2$

3) $x \geq 0$

4) $x < -3$

5) $x < \frac{1}{2}$

6) $x \leq -2$

7) $x \leq 3$

8) $x \geq -\frac{7}{2}$

WWW.MathNotion.Com

# ACT Aspire Subject Test Mathematics Grade 8

## One–Step Inequalities

✏️ **Find the answer for each inequality and graph it.**

1) $x + 4 \geq 4$

2) $x - 5 \leq 2$

3) $5x > 35$

4) $9 + x \leq 11$

5) $x - 5 < -9$

6) $9x \geq 72$

7) $9x \leq 27$

8) $x + 19 > 16$

# ACT Aspire Subject Test Mathematics Grade 8

## Multi-Step Inequalities

✎ **Calculate each inequality.**

1) $x - 3 \leq 7$

2) $8 - x \leq 8$

3) $3x - 9 \leq 9$

4) $4x - 4 \geq 8$

5) $x - 7 \geq 1$

6) $5x - 15 \leq 5$

7) $6x - 8 \leq 4$

8) $-11 + 6x \leq 12$

9) $4(x - 4) \leq 16$

10) $3x - 10 \leq 11$

11) $5x - 25 < 25$

12) $9x - 5 < 22$

13) $20 - 7x \geq -15$

14) $33 + 6x < 45$

15) $8 + 8x \geq 96$

16) $7 + 3x < 13$

17) $4x - 3 < 9$

18) $5(2 - 2x) \geq -30$

19) $-(7 + 6x) < 29$

20) $12 - 8x \geq -20$

21) $-4(x - 6) > 24$

22) $\dfrac{3x + 9}{6} \leq 10$

23) $\dfrac{4x - 10}{3} \leq 2$

24) $\dfrac{2x - 8}{3} > 2$

25) $8 + \dfrac{x}{6} < 9$

26) $\dfrac{9x}{7} - 4 < 5$

27) $\dfrac{15x + 45}{15} > 1$

28) $16 + \dfrac{x}{4} < 6$

WWW.MathNotion.Com

ACT Aspire Subject Test Mathematics Grade 8

## Systems of Equations

✏️ **Calculate each system of equations.**

1) $-x + y = 2$          $x = $ ___
   $-4x + 2y = 6$       $y = $ ___

2) $-15x + 3y = -9$   $x = $ ___
   $9x - 16y = 48$      $y = $ ___

3) $y = -7$                 $x = $ ___
   $6x + 5y = 7$          $y = $ ___

4) $3y = -9x + 15$    $x = $ ___
   $5x - 4y = -3$        $y = $ ___

5) $10x - 9y = -13$   $x = $ ___
   $-5x + 3y = 11$      $y = $ ___

6) $-12x - 16y = 20$  $x = $ ___
   $6x - 12y = 30$       $y = $ ___

7) $5x - 14y = -23$   $x = $ ___
   $-18x + 21y = 24$  $y = $ ___

8) $15x - 21y = -6$   $x = $ ___
   $2x - 3y = -2$         $y = $ ___

9) $-x + 3y = 3$         $x = $ ___
   $-14x + 16y = -10$ $y = $ ___

10) $x + 5y = 50$       $x = $ ___
    $3x + 10y = 80$     $y = $ ___

11) $6x - 7y = -8$      $x = $ ___
    $-x - 4y = -9$        $y = $ ___

12) $2x + 4y = -10$   $x = $ ___
    $2x - 8y = 14$        $y = $ ___

13) $4x + 3y = 12$     $x = $ ___
    $5x - 3y = 15$        $y = $ ___

14) $3x - 2y = 3$        $x = $ ___
    $7x - 8y = 22$        $y = $ ___

15) $3x + 2y = 5$       $x = $ ___
    $-10x - 4y = -14$  $y = $ ___

16) $10x + 7y = 1$     $x = $ ___
    $-5x - 7y = 24$      $y = $ ___

WWW.MathNotion.Com

ACT Aspire Subject Test Mathematics Grade 8

# Systems of Equations Word Problems

✎ **Find the answer for each word problem.**

1) Tickets to a movie cost $4 for adults and $3 for students. A group of friends purchased 8 tickets for $31.00. How many adults ticket did they buy? ____

2) At a store, Eva bought two shirts and five hats for $77.00. Nicole bought three same shirts and four same hats for $84.00. What is the price of each shirt? ____

3) A farmhouse shelters 18 animals, some are pigs, and some are ducks. Altogether there are 66 legs. How many pigs are there? ____

4) A class of 214 students went on a field trip. They took 36 vehicles, some cars and some buses. If each car holds 5 students and each bus hold 22 students, how many buses did they take? ____

5) A theater is selling tickets for a performance. Mr. Smith purchased 5 senior tickets and 3 child tickets for $105 for his friends and family. Mr. Jackson purchased 3 senior tickets and 5 child tickets for $79. What is the price of a senior ticket? $____

6) The difference of two numbers is 10. Their sum is 20. What is the bigger number? $____

7) The sum of the digits of a certain two-digit number is 7. Reversing its digits increase the number by 9. What is the number? ____

8) The difference of two numbers is 11. Their sum is 25. What are the numbers? ____

9) The length of a rectangle is 5 meters greater than 2 times the width. The perimeter of rectangle is 28 meters. What is the length of the rectangle? ____

10) Jim has 25 nickels and dimes totaling $1.80. How many nickels does he have? ____

# ACT Aspire Subject Test Mathematics Grade 8

## Answers of Worksheets

**One–Step Equations**

1) 30
2) 7
3) 4
4) 6
5) 5
6) 11
7) 11
8) 8
9) 17
10) −4
11) 12
12) 16
13) 34
14) −15
15) −18
16) 14
17) −14
18) 20
19) 45
20) −13
21) 34
22) −5
23) 47
24) −42
25) 22
26) 9
27) 60
28) 35
29) 54
30) 24
31) 30
32) −21

**Multi–Step Equations**

1) 2
2) −7
3) 4
4) 8
5) 6
6) 2
7) 10
8) 2
9) 2
10) −4
11) −4
12) −5
13) −7
14) 3
15) −2
16) 15
17) −2
18) −7
19) −5
20) −11
21) −4
22) −2
23) −7
24) 6
25) 3
26) 14
27) 7
28) 27
29) −5
30) 17
31) −6
32) −4

**Graphing Single–Variable Inequalities**

1)
2)
3)
4)

WWW.MathNotion.Com

# ACT Aspire Subject Test Mathematics Grade 8

5) [number line with open circle at 1/2]

6) [number line with closed circle at -2]

7) [number line with closed circle at 3]

8) [number line with closed circle at -3.5]

**One–Step Inequalities**

1) [number line with closed circle at 0]

2) [number line with closed circle at 7]

3) [number line with open circle at 7]

4) [number line with closed circle at 2]

5) [number line with open circle at -4]

6) [number line with closed circle at 8]

7) [number line with closed circle at 3]

8) [number line with open circle at -3]

**Multi-Step Inequalities**

1) $x \leq 10$
2) $x \geq 0$
3) $x \leq 6$
4) $x \geq 3$
5) $x \geq 8$
6) $x \leq 4$
7) $x \leq 2$
8) $x \leq \frac{23}{6}$
9) $x \leq 8$
10) $x \leq 7$
11) $x < 10$
12) $x < 3$
13) $x \leq 5$
14) $x < 2$
15) $x \geq 11$
16) $x < 2$
17) $x < 3$
18) $x \leq 4$
19) $x > -6$
20) $x \leq 4$
21) $x < 0$
22) $x \leq 17$
23) $x \leq 4$
24) $x > 7$

WWW.MathNotion.Com

# ACT Aspire Subject Test Mathematics Grade 8

25) $x < 6$   26) $x < 7$   27) $x > -2$   28) $x < -40$

**Systems of Equations**

1) $x = -1, y = 1$
2) $x = 0, y = -3$
3) $x = 7$
4) $x = 1, y = 2$
5) $x = -4, y = -3$
6) $x = 1, y = -2$
7) $x = 1, y = 2$
8) $x = 8, y = 6$
9) $x = 3, y = 2$
10) $x = -20, y = 14$
11) $x = 1, y = 2$
12) $x = -1, y = -2$
13) $x = 3, y = 0$
14) $x = -2, y = -\frac{9}{2}$
15) $x = 1, y = 1$
16) $x = 5, y = -7$

**Systems of Equations Word Problems**

1) 7
2) $16
3) 15
4) 2
5) $18
6) 15
7) 34
8) 18, 7
9) 11 meters
10) 14

ACT Aspire Subject Test Mathematics Grade 8

# Chapter 7 :
# Linear Functions

**Topics that you will practice in this chapter:**

- ✓ Finding Slope
- ✓ Graphing Lines Using Line Equation
- ✓ Writing Linear Equations
- ✓ Graphing Linear Inequalities
- ✓ Finding Midpoint
- ✓ Finding Distance of Two Points

*"Nature is written in mathematical language."* – Galileo Galilei

ACT Aspire Subject Test Mathematics Grade 8

# Finding Slope

🖎 **Find the slope of each line.**

1) $y = x + 8$

2) $y = -3x + 5$

3) $y = 2x + 12$

4) $y = -4x + 19$

5) $y = 11 + 6x$

6) $y = 7 - 5x$

7) $y = 8x + 19$

8) $y = -9x + 20$

9) $y = -7x + 4$

10) $y = 3x - 8$

11) $y = \frac{1}{3}x + 8$

12) $y = -\frac{4}{5}x + 9$

13) $-3x + 6y = 30$

14) $4x + 4y = 16$

15) $3y - x = 10$

16) $8y - x = 5$

🖎 **Find the slope of the line through each pair of points.**

17) $(2, 3), (7, 10)$

18) $(-3, 5), (2, 15)$

19) $(5, -3), (1, 9)$

20) $(-5, -5), (10, 25)$

21) $(22, 3), (7, 18)$

22) $(-16, 8), (-7, 26)$

23) $(25, 11), (29, 19)$

24) $(26, -19), (14, 17)$

25) $(22, -13), (20, -11)$

26) $(19, 7), (15, -3)$

27) $(5, 7), (11, 19)$

28) $(52, -62), (40, 70)$

WWW.MathNotion.Com

ACT Aspire Subject Test Mathematics Grade 8

## Graphing Lines Using Line Equation

✏️ **Sketch the graph of each line.**

1) $y = x - 2$

2) $y = -3x + 2$

3) $x + y = 0$

  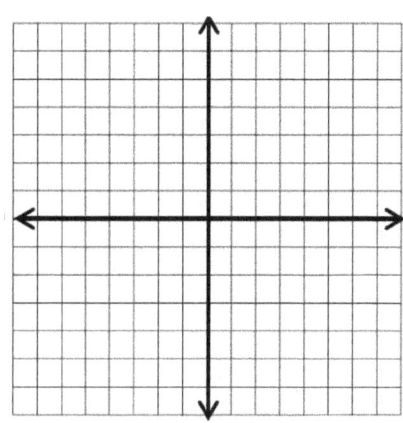

4) $x + y = -3$

5) $2x + 3y = -4$

6) $y - 3x + 6 = 0$

  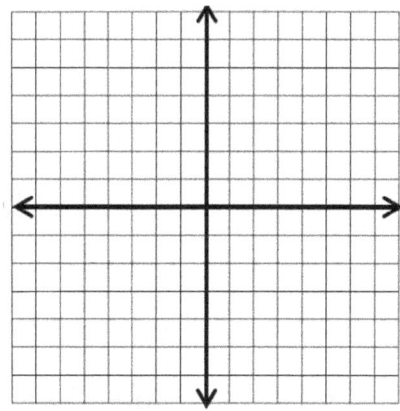

ACT Aspire Subject Test Mathematics Grade 8

## Writing Linear Equations

✎ **Write the equation of the line through the given points.**

1) Through: $(2, -5), (3, 9)$

2) Through: $(-6, 3), (3, 12)$

3) Through: $(10, 7), (5, 27)$

4) Through: $(15, 11), (3, -1)$

5) Through: $(24, 17), (12, -7)$

6) Through: $(8, 29), (4, -7)$

7) Through: $(20, -16), (12, 0)$

8) Through: $(-3, 10), (2, -5)$

9) Through: $(-6, 17), (4, -3)$

10) Through: $(-8, 22), (5, -4)$

11) Through: $(9, 27), (3, -3)$

12) Through: $(11, 32), (9, 4)$

13) Through: $(-3, 13), (-4, 0)$

14) Through: $(-5, 5), (5, 15)$

15) Through: $(18, -32), (11, 3)$

16) Through: $(-4, 25), (4, -15)$

✎ **Find the answer for each problem.**

17) What is the equation of a line with slope 6 and intercept 12? _____

18) What is the equation of a line with slope −11 and intercept −4? _____

19) What is the equation of a line with slope −3 and passes through point $(5, 2)$? _____

20) What is the equation of a line with slope −5 and passes through point $(-2, -1)$? _____

21) The slope of a line is −10 and it passes through point $(-3, 0)$. What is the equation of the line? _____

22) The slope of a line is 8 and it passes through point $(0, 7)$. What is the equation of the line? _____

WWW.MathNotion.Com

# ACT Aspire Subject Test Mathematics Grade 8

## Graphing Linear Inequalities

✏ **Sketch the graph of each linear inequality.**

1) $y > 4x - 5$    2) $y < 2x + 4$    3) $y \leq -5x - 2$

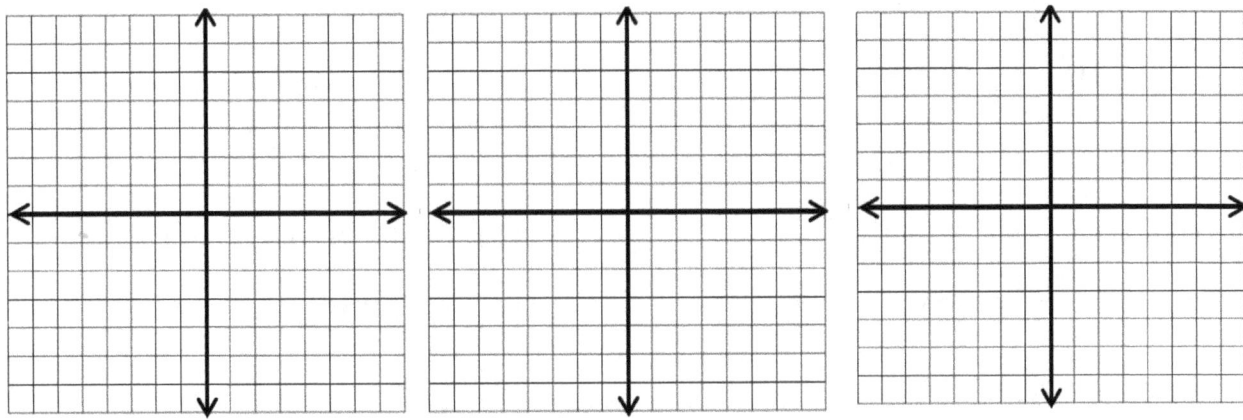

4) $4y \geq 12 + 4x$    5) $-12y < 3x - 24$    6) $5y \geq -15x + 10$

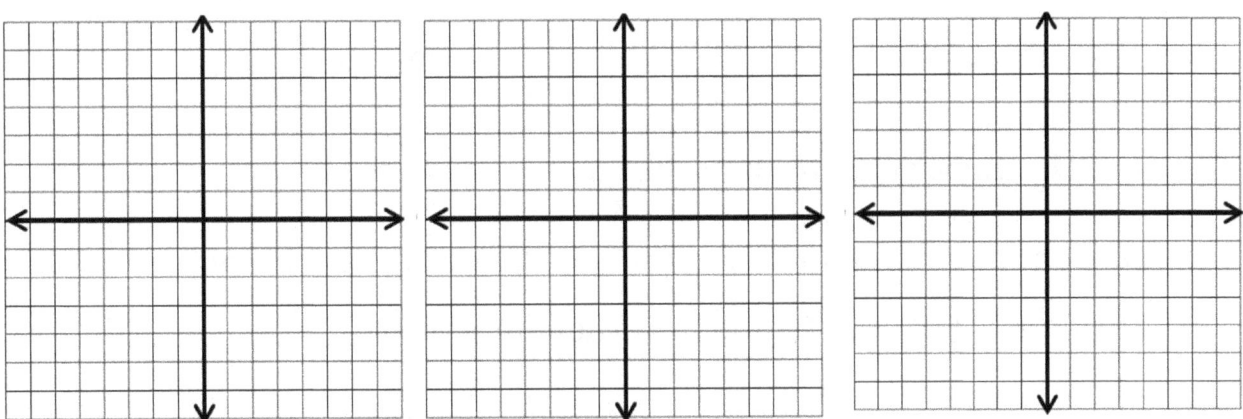

WWW.MathNotion.Com

## ACT Aspire Subject Test Mathematics Grade 8

## Finding Midpoint

✎ **Find the midpoint of the line segment with the given endpoints.**

1) $(-4, -3), (2, 3)$

2) $(9, 0), (-1, 8)$

3) $(9, -6), (3, 14)$

4) $(-10, -6), (0, 8)$

5) $(2, -5), (14, -15)$

6) $(-10, -3), (4, -13)$

7) $(8, 7), (-8, 13)$

8) $(-3, 6), (-9, 2)$

9) $(-4, 5), (16, -9)$

10) $(7, 14), (9, -2)$

11) $(-8, 6), (6, 6)$

12) $(10, 5), (-2, -3)$

13) $(-5, 12), (-3, 3)$

14) $(12, 7), (8, -2)$

15) $(10, 2), (-6, 14)$

16) $(-1, -2), (-7, 10)$

17) $(7, -7), (13, -13)$

18) $(-3, -8), (11, -4)$

19) $(5, -11), (-8, 9)$

20) $(14, -4), (16, 14)$

21) $(0, -5), (8, -1)$

22) $(3, 0), (-21, 18)$

23) $(17, -3), (-7, -5)$

24) $(26, -12), (6, 24)$

✎ **Find the answer for each problem.**

25) One endpoint of a line segment is $(-3, 7)$ and the midpoint of the line segment is $(-6, 9)$. What is the other endpoint? _____

26) One endpoint of a line segment is $(-3, 7)$ and the midpoint of the line segment is $(1, 5)$. What is the other endpoint? _____

27) One endpoint of a line segment is $(-10, -16)$ and the midpoint of the line segment is $(2, 9)$. What is the other endpoint? _____

## ACT Aspire Subject Test Mathematics Grade 8

## Finding Distance of Two Points

✎ Find the distance between each pair of points.

1) $(6, 3), (-3, -9)$

2) $(5, 2), (-10, -6)$

3) $(8, 5), (8, 3)$

4) $(-8, -2), (2, 22)$

5) $(6, -7), (-3, -7)$

6) $(12, 0), (-9, -20)$

7) $(3, 20), (3, -5)$

8) $(10, 17), (5, 5)$

9) $(7, -2), (-4, -2)$

10) $(13, 4), (5, -2)$

11) $(11, 13), (5, 5)$

12) $(1, 4), (-23, -3)$

13) $(9, 8), (5, -4)$

14) $(-11, -4), (5, 8)$

15) $(-2, -6), (-2, -12)$

16) $(-1, -4), (23, 3)$

17) $(19, 3), (7, -6)$

18) $(-5, -2), (3, 4)$

19) $(2, 6), (2, -12)$

20) $(-4, -2), (8, -2)$

✎ Find the answer for each problem.

21) Triangle ABC is a right triangle on the coordinate system and its vertices are $(-2, 5)$, $(-2, 1)$, and $(1, 1)$. What is the area of triangle ABC? _____

22) Three vertices of a triangle on a coordinate system are $(3, -6)$, $(-5, -12)$, and $(3, -18)$. What is the perimeter of the triangle? _____

23) Four vertices of a rectangle on a coordinate system are $(-2, 2)$, $(-2, 6)$, $(4, 2)$, and $(4, 6)$. What is its perimeter? _____

# ACT Aspire Subject Test Mathematics Grade 8

## Answers of Worksheets

**Finding Slope**

1) 1
2) −3
3) 2
4) −4
5) 6
6) −5
7) 8
8) −9
9) −7
10) 3
11) $\frac{1}{3}$
12) $-\frac{4}{5}$
13) $\frac{1}{2}$
14) −1
15) $\frac{1}{3}$
16) $\frac{1}{8}$
17) $\frac{7}{5}$
18) 2
19) −3
20) 2
21) −1
22) 2
23) 2
24) −3
25) −1
26) $\frac{5}{2}$
27) 2
28) −11

**Graphing Lines Using Line Equation**

1) $y = x - 2$

2) $y = -3x + 2$

3) $x + y = 0$

4) $x + y = -3$

5) $2x + 3y = -4$

6) $y - 3x + 6 = 0$

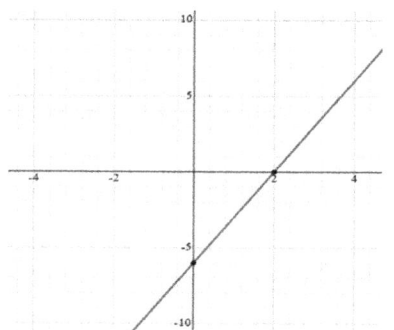

# ACT Aspire Subject Test Mathematics Grade 8

**Writing Linear Equations**

1) $y = 14x - 33$
2) $y = x + 9$
3) $y = -4x + 47$
4) $y = x - 4$
5) $y = 2x - 31$
6) $y = 9x - 43$
7) $y = -2x + 24$
8) $y = -3x + 1$
9) $y = -2x + 5$
10) $y = -2x + 6$
11) $y = 5x - 18$
12) $y = 14x - 122$
13) $y = 13x + 52$
14) $y = x + 10$
15) $y = -5x + 58$
16) $y = -5x + 5$
17) $y = 6x + 12$
18) $y = -11x - 4$
19) $y = -3x + 17$
20) $y = -5x - 11$
21) $y = -10x - 30$
22) $y = 8x + 7$

**Graphing Linear Inequalities**

1) $y > 4x - 5$

2) $y < 2x + 4$

3) $y \leq -5x - 2$

4) $4y \geq 12 + 4x$

5) $-12y < 3x - 24$

6) $5y \geq -15x + 10$

**Finding Midpoint**

1) $(-1, 0)$
2) $(4, 4)$
3) $(6, 4)$
4) $(-5, 1)$
5) $(8, -10)$
6) $(-3, -8)$
7) $(0, 10)$
8) $(-6, 4)$
9) $(6, -2)$
10) $(8, 6)$
11) $(-1, 6)$
12) $(4, 1)$
13) $(-4, 7.5)$
14) $(10, 2.5)$
15) $(2, 8)$
16) $(-4, 4)$
17) $(10, -10)$
18) $(4, -6)$

# ACT Aspire Subject Test Mathematics Grade 8

19) $(-1.5, -1)$
20) $(15, 5)$
21) $(4, -3)$

22) $(-9, 9)$
23) $(5, -4)$
24) $(16, 6)$

25) $(-9, 11)$
26) $(5, 3)$
27) $(14, 34)$

**Finding Distance of Two Points**

1) 15
2) 17
3) 2
4) 26
5) 9
6) 29
7) 25
8) 13

9) 11
10) 10
11) 10
12) 25
13) $4\sqrt{10}$
14) 20
15) 6
16) 25

17) 15
18) 10
19) 18
20) 12
21) 6 *square units*
22) 32 *units*
23) 20 *units*

# Chapter 8:
# Polynomials

**Topics that you will practice in this chapter:**

- ✓ Writing Polynomials in Standard Form
- ✓ Simplifying Polynomials
- ✓ Adding and Subtracting Polynomials
- ✓ Multiplying Monomials
- ✓ Multiplying and Dividing Monomials
- ✓ Multiplying a Polynomial and a Monomial
- ✓ Multiplying Binomials
- ✓ Factoring Trinomials
- ✓ Operations with Polynomials

*Mathematics is the supreme judge; from its decisions there is no appeal. – Tobias Dantzig*

ACT Aspire Subject Test Mathematics Grade 8

## Writing Polynomials in Standard Form

✎ Write each polynomial in standard form.

1) $11x - 7x =$

2) $-5 + 19x - 19x =$

3) $6x^5 - 12x^3 =$

4) $12 + 17x^4 - 12 =$

5) $5x^2 + 4x - 9x^3 =$

6) $-3x^2 + 12x^5 =$

7) $5x + 8x^3 - 2x^8 =$

8) $-7x^3 + 4x - 9x^6 =$

9) $3x^2 + 22 - 6x =$

10) $3 - 4x + 9x^4 =$

11) $13x^2 + 28x - 8x^3 =$

12) $16 + 4x^2 - 2x^3 =$

13) $19x^2 - 9x + 9x^4 =$

14) $3x^4 - 7x^2 - 2x^3 =$

15) $-51 + 3x^2 - 8x^4 =$

16) $7x^2 - 8x^6 + 4x^4 - 15 =$

17) $6x^4 - 4x^5 + 16 - 3x^3 =$

18) $-2x^6 + 4x - 7x^2 - 5x =$

19) $11x^7 + 8x^5 - 5x^7 - 3x^2 =$

20) $2x^2 - 12x^5 + 8x^2 + 3x^6 =$

21) $4x^5 - 11x^7 - 6x^3 + 16x^5 =$

22) $6x^3 + 3x^5 + 34x^4 - 8x^5 =$

23) $3x(4x + 5 - 2x^2) =$

24) $12x(x^6 + 4x^3) =$

25) $5x(3x^2 + 6x + 4) =$

26) $7x(4 - 2x + 6x^5) =$

27) $3x(4x^4 - 4x^3 + 2) =$

28) $4x(2x^5 + 6x^2 - 3) =$

29) $5x(3x^4 + 4x^3 + 2x) =$

30) $2x(3x - 2x^3 + 4x^6) =$

WWW.MathNotion.Com

# ACT Aspire Subject Test Mathematics Grade 8

## Simplifying Polynomials

✎ **Simplify each expression.**

1) $3(4x - 20) =$

2) $5x(3x - 4) =$

3) $6x(5x - 7) =$

4) $3x(7x + 5) =$

5) $5x(4x - 3) =$

6) $6x(8x + 2) =$

7) $(3x - 2)(x - 4) =$

8) $(x - 5)(2x + 6) =$

9) $(x - 3)(x - 7) =$

10) $(3x + 4)(3x - 4) =$

11) $(5x - 4)(5x - 2) =$

12) $6x^2 + 6x^2 - 8x^4 =$

13) $3x - 2x^2 + 5x^3 + 7 =$

14) $7x + 4x^2 - 10x^3 =$

15) $12x^2 + 5x^5 - 6x^3 =$

16) $-5x^2 + 4x^6 + 6x^8 =$

17) $-12x^3 + 10x^5 - 4x^6 + 4x =$

18) $11 - 7x^2 + 4x^2 - 16x^3 + 11 =$

19) $2x^2 - 9x + 4x^3 + 15x - 10x =$

20) $13 - 7x^5 + 6x^5 - 4x^2 + 5 =$

21) $-5x^8 + x^6 - 14x^3 + 5x^8 =$

22) $(7x^4 - 4) + (7x^4 - 2x^4) =$

23) $3(3x^4 - 4x^3 - 6x^4) =$

24) $-5(x^9 + 8) - 5(10 - x^9) =$

25) $8x^3 - 9x^4 - 2x + 19 - 8x^3 =$

26) $11 - 8x^3 + 6x^3 - 7x^5 + 6 =$

27) $(5x^3 - 4x) - (6x - 2 - 6x^3) =$

28) $4x^2 - 5x^4 - x(3x^3 + 2x) =$

29) $6x + 6x^5 - 10 - 4(x^5 - 3) =$

30) $4 - 3x^4 + (6x^5 - 2x^4 + 5x^5) =$

31) $-(x^5 + 4) - 8(3 + x^5) =$

32) $(4x^3 - 3x) - (3x - 5x^3) =$

WWW.MathNotion.Com

ACT Aspire Subject Test Mathematics Grade 8

## Adding and Subtracting Polynomials

✏ **Add or subtract expressions.**

1) $(-2x^2 - 3) + (3x^2 + 4) =$

2) $(4x^3 + 6) - (7 - 2x^3) =$

3) $(4x^5 + 5x^2) - (2x^5 + 15) =$

4) $(6x^3 - 2x^2) + (5x^2 - 4x) =$

5) $(10x^4 + 28x) - (34x^4 + 6) =$

6) $(7x^2 - 3) + (7x^2 + 3) =$

7) $(9x^2 + 4) - (10 - 5x^2) =$

8) $(6x^2 + x^5) - (x^5 + 4) =$

9) $(4x^3 - x) + (3x - 7x^3) =$

10) $(11x + 10) - (8x + 10) =$

11) $(15x^3 - 3x) - (3x - 4x^3) =$

12) $(4x - x^5) - (6x^5 + 8x) =$

13) $(2x^2 - 7x^7) - (4x^7 - 6x) =$

14) $(3x^2 - 5) + (8x^2 + 4x^5) =$

15) $(9x^4 + 5x^5) - (x^5 - 9x^4) =$

16) $(-4x^3 - 2x) + (9x - 5x^3) =$

17) $(4x - 3x^2) - (148x^2 + x) =$

18) $(5x - 8x^4) - (3x^4 - 4x^2) =$

19) $(8x^4 - 4) + (2x^4 - 3x^2) =$

20) $(5x^6 + 7x^3) - (x^3 - 5x^6) =$

21) $(-2x^2 + 20x^5 + 5x^4) + (12x^4 + 8x^5 + 24x^2) =$

22) $(7x^4 - 9x^7 - 6x) - (-3x^4 - 9x^7 + 6x) =$

23) $(14x + 12x^4 - 18x^6) + (20x^4 + 18x^6 - 10x) =$

24) $(5x^8 - 6x^6 - 4x) - (5x^3 + 9x^6 - 7x) =$

25) $(11x^2 - 6x^4 - 3x) - (-4x^2 - 12x^4 + 9x) =$

26) $(-5x^9 + 14x^3 + 3x^7) + (10x^7 + 26x^3 + 3x^9) =$

WWW.MathNotion.Com

ACT Aspire Subject Test Mathematics Grade 8

## Multiplying Monomials

✏️ **Simplify each expression.**

1) $6u^8 \times (-u^2) =$

2) $(-5p^8) \times (-2p^3) =$

3) $4xy^3z^5 \times 3z^4 =$

4) $3u^5t \times 8ut^4 =$

5) $(-5a^2) \times (-7a^3b^6) =$

6) $-3a^4b^3 \times 6a^2b =$

7) $13xy^5 \times x^4y^4 =$

8) $6p^4q^3 \times (-8pq^6) =$

9) $8s^4t^3 \times 4st^3 =$

10) $(-6x^4y^3) \times 6x^2y =$

11) $3xy^7z \times 12z^3 =$

12) $24xy \times x^2y =$

13) $13pq^4 \times (-3p^2q) =$

14) $13s^3t^4 \times st^4 =$

15) $11p^5 \times (-6p^3) =$

16) $(-8p^3q^5r) \times 3pq^4r^6 =$

17) $(-4a^4) \times (-7a^3b) =$

18) $6u^6v^2 \times (-5u^3v^4) =$

19) $9u^5 \times (-3u) =$

20) $-6xy^5 \times 4x^2y =$

21) $13y^5z^3 \times (-y^3z) =$

22) $8a^4bc^3 \times 2abc^3 =$

23) $(-7p^5q^6) \times (-5p^4q^2) =$

24) $4u^5v^3 \times (-4u^7v^3) =$

25) $17y^4z^5 \times (-y^6z) =$

26) $(-5pq^3r^2) \times 8p^2q^4r =$

27) $3ab^5c^6 \times 5a^4bc^2 =$

28) $6x^3yz^2 \times 3x^2y^7z^3 =$

WWW.MathNotion.Com

**ACT Aspire Subject Test Mathematics Grade 8**

## Multiplying and Dividing Monomials

✏️ Simplify each expression.

1) $(5x^5)(2x^2) =$

2) $(4x^4)(6x^2) =$

3) $(3x^4)(7x^4) =$

4) $(5x^6)(4x^2) =$

5) $(12x^4)(3x^6) =$

6) $(4yx^8)(8y^4x^3) =$

7) $(14x^4y)(x^3y^5) =$

8) $(-5x^3y^4)(2x^3y^5) =$

9) $(-6x^4y^2)(-3x^3y^5) =$

10) $(5x^3y)(-5x^2y^3) =$

11) $(6x^4y^3)(4x^3y^4) =$

12) $(4x^3y^2)(5x^2y^4) =$

13) $(12x^3y^6)(4x^4y^{10}) =$

14) $(15x^3y^5)(3x^4y^6) =$

15) $(7x^2y^7)(8x^6y^7) =$

16) $(-3x^3y^8)(7x^9y^4) =$

17) $\dfrac{5x^6y^6}{xy^4} =$

18) $\dfrac{19x^7y^5}{19x^6y} =$

19) $\dfrac{56x^4y^4}{8xy} =$

20) $\dfrac{81x^5y^6}{9x^4y^5} =$

21) $\dfrac{36x^7y^6}{9x^2y^3} =$

22) $\dfrac{48x^9y^7}{4x^4y^6} =$

23) $\dfrac{88x^{18}y^{12}}{11x^8y^9} =$

24) $\dfrac{30x^7y^6}{6x^8y^3} =$

25) $\dfrac{150x^7y^6}{30x^4y^6} =$

26) $\dfrac{-42x^{18}y^{14}}{6x^4y^9} =$

27) $\dfrac{-36x^7y^8}{9x^5y^8} =$

WWW.MathNotion.Com

# Multiplying a Polynomial and a Monomial

✏ Find each product.

1) $x(2x + 4) =$

2) $6(4 - 2x) =$

3) $5x(4x + 2) =$

4) $x(-4x + 5) =$

5) $8x(2x - 2) =$

6) $6(2x - 4y) =$

7) $7x(5x - 5) =$

8) $3x(12x + 2y) =$

9) $4x(x + 6y) =$

10) $11x(3x + 4y) =$

11) $7x(3x + 2) =$

12) $10x(4x - 10y) =$

13) $9x(3x - 2y) =$

14) $7x(x - 4y + 6) =$

15) $8x(2x^2 + 5y^2) =$

16) $12x(2x + 3y) =$

17) $4(2x^4 - 4y^4) =$

18) $4x(-3x^2y + 4y) =$

19) $-4(5x^3 - 2xy + 4) =$

20) $4(x^2 - 5xy - 6) =$

21) $8x(2x^3 - 5xy + 2x) =$

22) $-6x(-2x^3 - 6x + 2xy) =$

23) $3(2x^2 + xy - 9y^2) =$

24) $4x(5x^3 - 3x + 7) =$

25) $6(3x^{22} - 2x - 5) =$

26) $x^2(-2x^3 + 4x + 3) =$

27) $x^2(4x^3 + 10 - 2x) =$

28) $4x^4(3x^3 - 2x + 5) =$

29) $2x^2(4x^4 - 5xy + 7y^3) =$

30) $5x^2(5x^4 - 3x + 9) =$

31) $7x^2(6x^2 + 3x - 6) =$

32) $4x(x^3 - 4xy + 2y^2) =$

## Multiplying Binomials

✏ **Find each product.**

1) $(x + 3)(x + 6) =$

2) $(x - 4)(x + 3) =$

3) $(x - 3)(x - 8) =$

4) $(x + 8)(x + 9) =$

5) $(x - 2)(x - 12) =$

6) $(x + 5)(x + 5) =$

7) $(x - 6)(x + 7) =$

8) $(x - 8)(x - 3) =$

9) $(x + 7)(x + 12) =$

10) $(x - 4)(x + 8) =$

11) $(x + 8)(x + 8) =$

12) $(x + 2)(x + 7) =$

13) $(x - 6)(x + 6) =$

14) $(x - 5)(x + 5) =$

15) $(x + 11)(x + 11) =$

16) $(x + 6)(x + 9) =$

17) $(x - 2)(x + 2) =$

18) $(x - 4)(x + 7) =$

19) $(3x + 5)(x + 6) =$

20) $(5x - 6)(4x + 8) =$

21) $(x - 7)(3x + 7) =$

22) $(x - 9)(x - 4) =$

23) $(x - 12)(x + 2) =$

24) $(2x - 4)(5x + 4) =$

25) $(3x - 8)(x + 8) =$

26) $(7x - 2)(6x + 3) =$

27) $(4x + 5)(3x + 5) =$

28) $(7x - 4)(9x + 4) =$

29) $(x + 2)(2x - 8) =$

30) $(5x - 4)(5x + 4) =$

31) $(3x + 2)(3x - 7) =$

32) $(x^2 + 8)(x^2 - 8) =$

ACT Aspire Subject Test Mathematics Grade 8

# Factoring Trinomials

✎ **Factor each trinomial.**

1) $x^2 + 8x + 12 =$

2) $x^2 - 6x + 5 =$

3) $x^2 + 15x + 36 =$

4) $x^2 - 12x + 35 =$

5) $x^2 - 11x + 18 =$

6) $x^2 - 9x + 18 =$

7) $x^2 + 18x + 72 =$

8) $x^2 - x - 72 =$

9) $x^2 + 4x - 21 =$

10) $x^2 - 13x + 22 =$

11) $x^2 + 2x - 24 =$

12) $x^2 - 3x - 40 =$

13) $x^2 - 3x - 70 =$

14) $x^2 + 26x + 169 =$

15) $4x^2 - 7x - 15 =$

16) $x^2 - 14x + 33 =$

17) $10x^2 + 5x - 15 =$

18) $6x^2 - 4x - 42 =$

19) $x^2 + 12x + 36 =$

20) $5x^2 + 17x - 12 =$

✎ **Calculate each problem.**

21) The area of a rectangle is $x^2 - x - 56$. If the width of rectangle is $x + 7$, what is its length? _____

22) The area of a parallelogram is $4x^2 + 17x - 15$ and its height is $x + 5$. What is the base of the parallelogram? _____

23) The area of a rectangle is $6x^2 - 22x + 12$. If the width of the rectangle is $3x - 2$, what is its length? _____

WWW.MathNotion.Com

ACT Aspire Subject Test Mathematics Grade 8

## Operations with Polynomials

✏ **Find each product.**

1) $4(5x + 3) = $ _____

2) $8(2x + 6) = $ _____

3) $2(5x - 2) = $ _____

4) $-4(7x - 3) = $ _____

5) $3x^2(9x + 1) = $ _____

6) $4x^6(7x - 9) = $ _____

7) $3x^4(-7x + 3) = $ _____

8) $-8x^4(5x - 8) = $ _____

9) $7(x^2 + 5x - 3) = $ _____

10) $9(5x^2 - 7x + 5) = $ _____

11) $3(3x^2 + 3x + 2) = $ _____

12) $5x(3x^2 + 5x + 8) = $ _____

13) $(5x + 7)(3x - 3) = $ _____

14) $(9x + 3)(3x - 5) = $ _____

15) $(6x + 3)(4x - 2) = $ _____

16) $(7x - 2)(3x + 5) = $ _____

✏ **Calculate each problem.**

17) The measures of two sides of a triangle are $(2x + 5y)$ and $(6x - 3y)$. If the perimeter of the triangle is $(13x + 4y)$, what is the measure of the third side? _____

18) The height of a triangle is $(8x + 5)$ and its base is $(4x - 3)$. What is the area of the triangle? _____

19) One side of a square is $(6x + 2)$. What is the area of the square? _____

20) The length of a rectangle is $(5x - 8y)$ and its width is $(15x + 8y)$. What is the perimeter of the rectangle? _____

21) The side of a cube measures $(x + 2)$. What is the volume of the cube? _____

22) If the perimeter of a rectangle is $(28x + 6y)$ and its width is $(5x + 2y)$, what is the length of the rectangle? _____

# ACT Aspire Subject Test Mathematics Grade 8

## Answers of Worksheets

**Writing Polynomials in Standard Form**

1) $4x$
2) $-5$
3) $6x^5 - 12x^3$
4) $14x^4$
5) $-9x^3 + 5x^2 + 4x$
6) $12x^5 - 3x^2$
7) $-2x^8 + 8x^3 + 5x$
8) $-9x^6 - 7x^3 + 4x$
9) $3x^2 - 6x + 22$
10) $9x^4 - 4x + 3$
11) $-8x^3 + 13x^2 + 28x$
12) $-2x^3 + 4x^2 + 16$
13) $9x^4 + 19x^2 - 9x$
14) $3x^4 - 2x^3 - 7x^2$
15) $-8x^4 + 3x^2 - 51$
16) $-8x^6 + 4x^4 + 7x^2 - 15$
17) $-4x^5 + 6x^4 - 3x^3 + 16$
18) $-2x^6 - 7x^2 - x$
19) $6x^7 + 8x^5 - 3x^2$
20) $3x^6 - 12x^5 + 10x^2$
21) $-11x^7 + 20x^5 - 6x^3$
22) $-5x^5 + 34x^4 + 6x^3$
23) $-6x^3 + 12x^2 + 15x$
24) $12x^7 + 48x^4$
25) $15x^3 + 30x^2 + 20x$
26) $42x^6 - 14x^2 + 28x$
27) $12x^5 - 12x^4 + 6x$
28) $8x^6 + 24x^3 - 12x$
29) $15x^5 + 20x^4 + 10x^2$
30) $8x^7 - 4x^4 + 6x^2$

**Simplifying Polynomials**

1) $12x - 60$
2) $15x^2 - 20x$
3) $30x^2 - 42x$
4) $21x^2 + 15x$
5) $20x^2 - 15x$
6) $48x^2 + 12x$
7) $3x^2 - 14x + 8$
8) $2x^2 - 4x - 30$
9) $x^2 - 10x + 21$
10) $9x^2 - 16$
11) $25x^2 - 30x + 8$
12) $-8x^4 + 12x^2$
13) $5x^3 - 2x^2 + 3x + 7$
14) $-10x^3 + 4x^2 + 7x$
15) $5x^5 - 6x^3 + 12x^2$
16) $6x^8 + 4x^6 - 5x^2$
17) $-4x^6 + 10x^5 - 12x^3 + 4x$
18) $-16x^3 - 3x^2 + 22$
19) $4x^3 + 2x^2 - 4x$
20) $-x^5 - 4x^2 + 18$
21) $x^6 - 14x^3$
22) $12x^4 - 4$
23) $-9x^4 - 12x^3$
24) $-90$

WWW.MathNotion.Com

# ACT Aspire Subject Test Mathematics Grade 8

25) $-9x^4 - 2x + 19$
26) $-7x^5 - 2x^3 + 17$
27) $11x^3 - 10x + 2$
28) $-8x^4 + 2x^2$

29) $2x^5 + 6x + 2$
30) $11x^5 - 5x^4 + 4$
31) $-9x^5 - 28$
32) $9x^3 - 6x$

**Adding and Subtracting Polynomials**

1) $x^2 + 1$
2) $6x^3 - 1$
3) $2x^5 + 5x^2 - 15$
4) $6x^3 + 3x^2 - 4x$
5) $-24x^4 + 28x - 6$
6) $14x^2$
7) $14x^2 - 6$
8) $6x^2 - 4$
9) $-3x^3 + 2x$

10) $3x$
11) $19x^3 - 6x$
12) $-7x^5 - 4x$
13) $-11x^7 + 2x^2 + 6x$
14) $4x^5 + 11x^2 - 5$
15) $4x^5 + 18x^4$
16) $-9x^3 + 7x$
17) $-151x^2 + 3x$
18) $-11x^4 + 4x^2 + 5x$

19) $10x^4 - 3x^2 - 4$
20) $10x^6 + 6x^3$
21) $28x^5 + 17x^4 + 22x^2$
22) $10x^4 - 12x$
23) $32x^4 + 4x$
24) $5x^8 - 15x^6 - 5x^3 + 3x$
25) $6x^4 + 15x^2 - 12x$
26) $-2x^9 + 13x^7 + 40x^3$

**Multiplying Monomials**

1) $-6u^{10}$
2) $10p^{11}$
3) $12xy^3z^9$
4) $24u^6t^5$
5) $35a^5b^6$
6) $-18a^6b^4$
7) $13x^5y^9$
8) $-48p^5q^9$
9) $32s^5t^6$
10) $-36x^6y^4$

11) $36xy^7z^4$
12) $24px^3y^2$
13) $-39p^3q^5$
14) $13s^4t^8$
15) $-66p^8$
16) $-24p^4q^9r^7$
17) $28a^7b$
18) $-30u^9v^6$
19) $-27u^6$
20) $-24x^3y^6$

21) $-13y^8z^4$
22) $16a^5b^2c^6$
23) $35p^9q^8$
24) $-16u^{12}v^6$
25) $-17y^{10}z^6$
26) $-40p^3q^7r^3$
27) $15a^5b^6c^8$
28) $18x^5y^8z^5$

**Multiplying and Dividing Monomials**

1) $10x^7$
2) $24x^6$
3) $21x^8$
4) $20x^8$

5) $36x^{10}$
6) $32x^{11}y^5$
7) $14x^7y^6$
8) $-10x^6y^9$

9) $18x^7y^7$
10) $-25x^5y^4$
11) $24x^7y^7$
12) $20x^5y^6$

WWW.MathNotion.Com

# ACT Aspire Subject Test Mathematics Grade 8

13) $48x^7y^{16}$
14) $45x^7y^{11}$
15) $56x^8y^{14}$
16) $-21x^{12}y^{12}$
17) $5x^5y^2$

18) $xy^4$
19) $7x^3y^3$
20) $9xy$
21) $4x^5y^3$
22) $12x^5y$

23) $8x^{10}y^3$
24) $5x^{-1}y^3$
25) $5x^3$
26) $-7x^{14}y^5$
27) $-4x^2$

**Multiplying a Polynomial and a Monomial**

1) $2x^2 + 4x$
2) $-12x + 24$
3) $20x^2 + 10x$
4) $-4x^2 + 5x$
5) $16x^2 - 16x$
6) $12x - 24y$
7) $35x^2 - 35x$
8) $36x^2 + 6xy$
9) $4x^2 + 24xy$
10) $33x^2 + 44xy$
11) $21x^2 + 14x$
12) $40x^2 - 100xy$
13) $27x^2 - 18xy$
14) $7x^2 - 28xy + 42x$
15) $16x^3 + 40xy^2$
16) $24x^2 + 36xy$

17) $8x^4 - 16y^4$
18) $-12x^3y + 16xy$
19) $-20x^3 + 8xy - 16$
20) $4x^2 - 20xy - 24$
21) $16x^4 - 40x^2y + 16x^2$
22) $12x^4 + 36x^2 - 12x^2y$
23) $6x^2 + 3xy - 27y^2$
24) $20x^4 - 12x^2 + 28x$
25) $18x^{22} - 12x - 30$
26) $-2x^5 + 4x^3 + 3x^2$
27) $4x^5 - 2x^3 + 10x^2$
28) $12x^7 - 8x^5 + 20x^4$
29) $8x^6 - 10x^3y + 14x^2y^3$
30) $25x^6 - 15x^3 + 45x^2$
31) $42x^4 + 21x^3 - 42x^2$
32) $4x^4 - 16x^2y + 8xy^2$

**Multiplying Binomials**

1) $x^2 + 9x + 18$
2) $x^2 - x - 12$
3) $x^2 - 11x + 24$
4) $x^2 + 17x + 72$
5) $x^2 - 14x + 24$
6) $x^2 + 10x + 25$
7) $x^2 + x - 42$

8) $x^2 - 11x + 24$
9) $x^2 + 19x + 84$
10) $x^2 + 4x - 32$
11) $x^2 + 16x + 64$
12) $x^2 + 9x + 14$
13) $x^2 - 36$
14) $x^2 - 25$

# ACT Aspire Subject Test Mathematics Grade 8

15) $x^2 + 22x + 121$
16) $x^2 + 15x + 54$
17) $x^2 - 4$
18) $x^2 + 3x - 28$
19) $3x^2 + 23x + 30$
20) $20x^2 + 16x - 48$
21) $3x^2 - 14x - 49$
22) $x^2 - 13x + 36$
23) $x^2 - 10x - 24$

24) $10x^2 - 12x - 16$
25) $3x^2 + 16x - 64$
26) $42x^2 + 9x - 6$
27) $12x^2 + 35x + 25$
28) $63x^2 - 8x - 16$
29) $2x^2 - 4x - 16$
30) $25x^2 - 16$
31) $9x^2 - 15x - 14$
32) $x^4 - 64$

## Factoring Trinomials

1) $(x + 6)(x + 2)$
2) $(x - 5)(x - 1)$
3) $(x + 12)(x + 3)$
4) $(x - 5)(x - 7)$
5) $(x - 2)(x - 9)$
6) $(x - 6)(x - 3)$
7) $(x + 6)(x + 12)$
8) $(x + 8)(x - 9)$

9) $(x - 3)(x + 7)$
10) $(x - 11)(x - 2)$
11) $(x - 4)(x + 6)$
12) $(x - 8)(x + 5)$
13) $(x + 7)(x - 10)$
14) $(x + 13)(x + 13)$
15) $(4x + 5)(x - 3)$
16) $(x - 11)(x - 3)$

17) $(5x - 5)(2x + 3)$
18) $(2x - 6)(3x + 7)$
19) $(x + 6)(x + 6)$
20) $(5x - 3)(x + 4)$
21) $(x - 8)$
22) $(4x - 3)$
23) $(2x - 6)$

## Operations with Polynomials

1) $20x + 12$
2) $16x + 48$
3) $10x - 4$
4) $-28x + 12$
5) $27x^3 + 3x^2$
6) $28x^7 - 36x^6$
7) $-21x^5 + 9x^4$
8) $-40x^5 + 64x^4$

9) $7x^2 + 35x - 21$
10) $45x^2 - 63x + 45$
11) $9x^2 + 9x + 6$
12) $15x^3 + 25x^2 + 40x$
13) $15x^2 + 6x - 21$
14) $27x^2 - 36x - 15$
15) $24x^2 - 6$
16) $21x^2 + 29x - 10$

17) $(5x + 2y)$
18) $16x^2 - 2x - \frac{15}{2}$
19) $36x^2 + 24x + 4$
20) $40x$
21) $x^3 + 6x^2 + 12x + 8$
22) $(9x + y)$

ACT Aspire Subject Test Mathematics Grade 8

# Chapter 9 :
# Functions Operations and Quadratic

**Topics that you will practice in this chapter:**

- ✓ Evaluating Function
- ✓ Adding and Subtracting Functions
- ✓ Multiplying and Dividing Functions
- ✓ Composition of Functions
- ✓ Quadratic Equation
- ✓ Solving Quadratic Equations
- ✓ Quadratic Formula and the Discriminant
- ✓ Quadratic Inequalities
- ✓ Graphing Quadratic Functions
- ✓ Domain and Range of Radical Functions
- ✓ Solving Radical Equations

*It's fine to work on any problem, so long as it generates interesting mathematics along the way – even if you don't solve it at the end of the day." – Andrew Wiles*

# ACT Aspire Subject Test Mathematics Grade 8

## Evaluating Function

### ✎ Write each of following in function notation.

1) $h = -8x + 3$

2) $k = 2a - 14$

3) $d = 11t$

4) $y = \frac{5}{12}x - \frac{7}{12}$

5) $m = 24n - 210$

6) $c = p^2 - 5p + 10$

### ✎ Evaluate each function.

7) $f(x) = 2x - 7$, find $f(-3)$

8) $g(x) = \frac{1}{9}x + 12$, find $f(18)$

9) $h(x) = -4x + 9$, find $f(3)$

10) $f(x) = -x + 19$, find $f(-3)$

11) $f(a) = 7a - 12$, find $f(3)$

12) $h(x) = 14 - 3x$, find $f(-4)$

13) $g(n) = 6n - 10$, find $f(2)$

14) $f(x) = -11x - 4$, find $f(-1)$

15) $k(n) = -20 - 3.5n$, find $f(2)$

16) $f(x) = -0.7x + 3.3$, find $f(-7)$

17) $g(n) = \frac{11n+8}{n}$, find $g(2)$

18) $g(n) = \sqrt{3n} + 12$, find $g(3)$

19) $h(x) = x^{-2} - 7$, find $h(\frac{1}{9})$

20) $h(n) = n^{-3} + 11$, find $h(\frac{1}{4})$

21) $h(n) = n^3 - 2$, find $h(\frac{1}{2})$

22) $h(n) = n^2 - 4$, find $h(-\frac{1}{3})$

23) $h(n) = 4n^2 - 13$, find $h(-5)$

24) $h(n) = -2n^3 - 6n$, find $h(2)$

25) $g(n) = \sqrt{16n^2} - \sqrt{n}$, find $g(4)$

26) $h(a) = \frac{-14a+9}{3a}$, find $h(-b)$

27) $k(a) = 12a - 14$, find $k(a - 3)$

28) $h(x) = \frac{1}{9}x + 18$, find $h(-18x)$

29) $h(x) = 8x^2 + 16$, find $h(\frac{x}{2})$

30) $h(x) = x^4 - 20$, find $h(-2x)$

WWW.MathNotion.Com

**ACT Aspire Subject Test Mathematics Grade 8**

## Adding and Subtracting Functions

✏️ **Perform the indicated operation.**

1) $f(x) = 2x + 3$

   $g(x) = x + 7$

   Find $(f - g)(2)$

2) $g(a) = -5a - 8$

   $f(a) = -3a - 5$

   Find $(g - f)(-2)$

3) $h(t) = 4t + 3$

   $g(t) = 4t + 7$

   Find $(h - g)(t)$

4) $g(a) = -6a - 10$

   $f(a) = 3a^2 + 9$

   Find $(g - f)(x)$

5) $g(x) = \frac{5}{6}x - 23$

   $h(x) = \frac{5}{12}x + 25$

   Find $g(12) - h(12)$

6) $h(x) = \sqrt{3x} - 2$

   $g(x) = \sqrt{3x} + 5$

   Find $(h + g)(12)$

7) $f(x) = x^{-1}$

   $g(x) = x^2 + \frac{5}{x}$

   Find $(f - g)(-3)$

8) $h(n) = n^2 + 2$

   $g(n) = -4n + 6$

   Find $(h - g)(2a)$

9) $g(x) = -2x^2 - 5 - 4x$

   $f(x) = 7 + 2x$

   Find $(g - f)(3x)$

10) $g(t) = 11t - 4$

    $f(t) = -2t^2 + 5$

    Find $(g + f)(-t)$

11) $f(x) = 8x + 9$

    $g(x) = -5x^2 + 3x$

    Find $(f - g)(-x^2)$

12) $f(x) = -3x^4 - 5x$

    $g(x) = 2x^4 + 5x + 22$

    Find $(f + g)(3x^2)$

WWW.MathNotion.Com

ACT Aspire Subject Test Mathematics Grade 8

## Multiplying and Dividing Functions

✎ **Perform the indicated operation.**

1) $g(x) = -2x - 1$
   $f(x) = 4x + 3$
   Find $(g.f)(2)$

2) $f(x) = 5x$
   $h(x) = -2x + 3$
   Find $(f.h)(-2)$

3) $g(a) = 5a - 2$
   $h(a) = 2a - 3$
   Find $(g.h)(-3)$

4) $f(x) = 2x - 7$
   $h(x) = x - 5$
   Find $(\frac{f}{h})(4)$

5) $f(x) = 8a^2$
   $g(x) = 3 + 2a$
   Find $(\frac{f}{g})(2)$

6) $g(a) = \sqrt{4a} + 2$
   $f(a) = (-a)^4 + 1$
   Find $(\frac{g}{f})(1)$

7) $g(t) = t^3 + 1$
   $h(t) = 5t - 2$
   Find $(g.h)(-2)$

8) $g(n) = n^2 + 2n - 4$
   $h(n) = -5n + 3$
   Find $(g.h)(1)$

9) $g(a) = (a - 3)^2$
   $f(a) = a^2 + 4$
   Find $(\frac{g}{f})(3)$

10) $g(x) = -3x^2 + \frac{4}{5}x + 9$
    $f(x) = x^2 - 24$
    Find $(\frac{g}{f})(5)$

11) $f(x) = 2x^3 - 5x^2 + 1$
    $g(x) = 3x - 1$
    Find $(f.g)(x)$

12) $f(x) = 5x - 2$
    $g(x) = x^3 - 2x$
    Find $(f.g)(x^2)$

WWW.MathNotion.Com

# ACT Aspire Subject Test Mathematics Grade 8

## Composition of Functions

Using $f(x) = 2x - 5$ and $g(x) = -2x$, find:

1) $f(g(2)) =$

2) $f(g(-1)) =$

3) $g(f(-4)) =$

4) $g(f(5)) =$

5) $f(g(3)) =$

6) $g(f(0)) =$

Using $f(x) = -\frac{1}{4}x + \frac{3}{4}$ and $g(x) = 2x^2$, find:

7) $g(f(-2)) =$

8) $g(f(4)) =$

9) $g(g(1)) =$

10) $f(f(1)) =$

11) $g(f(-4)) =$

12) $f(g(x)) =$

Using $f(x) = -2x + 2$ and $g(x) = x + 1$, find:

13) $g(f(1)) =$

14) $f(f(0)) =$

15) $f(g(-1)) =$

16) $f(g(-3)) =$

17) $g(f(2)) =$

18) $f(g(x)) =$

Using $f(x) = \sqrt{x + 9}$ and $g(x) = x - 9$, find:

19) $f(g(9)) =$

20) $g(f(-9)) =$

21) $f(g(4)) =$

22) $f(f(7)) =$

23) $g(f(-5)) =$

24) $g(g(0)) =$

ACT Aspire Subject Test Mathematics Grade 8

# Quadratic Equation

## ✎ Multiply.

1) $(x-4)(x+6) = $ _____

2) $(x+5)(x+7) = $ _____

3) $(x-6)(x+8) = $ _____

4) $(x+2)(x-9) = $ _____

5) $(x-7)(x-8) = $ _____

6) $(3x+2)(x-3) = $ _____

7) $(4x-3)(x+2) = $ _____

8) $(4x-5)(x+1) = $ _____

9) $(7x+1)(x-6) = $ _____

10) $(5x+1)(3x-3) = $ _____

## ✎ Factor each expression.

11) $x^2 - 2x - 8 = $ _____

12) $x^2 + 8x + 15 = $ _____

13) $x^2 - 2x - 24 = $ _____

14) $x^2 - 10x + 21 = $ _____

15) $x^2 + 10x + 21 = $ _____

16) $4x^2 + 9x + 5 = $ _____

17) $5x^2 + 13x - 6 = $ _____

18) $5x^2 + 17x - 12 = $ _____

19) $2x^2 + 7x + 5 = $ _____

20) $9x^2 - 21x + 6 = $ _____

## ✎ Calculate each equation.

21) $(x+6)(x-3) = 0$

22) $(x+1)(x+8) = 0$

23) $(3x+6)(x+5) = 0$

24) $(2x-2)(4x+8) = 0$

25) $x^2 + x + 10 = 22$

26) $x^2 + 11x + 36 = 12$

27) $2x^2 + 9x + 9 = 5$

28) $x^2 + 3x - 24 = 4$

29) $5x^2 + 5x - 40 = 20$

30) $8x^2 + 8x = 48$

WWW.MathNotion.Com

# ACT Aspire Subject Test Mathematics Grade 8

## Solving Quadratic Equations

✎ **Solve each equation by factoring or using the quadratic formula.**

1) $(x+9)(x-1) = 0$

2) $(x+7)(x+6) = 0$

3) $(x-8)(x+3) = 0$

4) $(x-6)(x-4) = 0$

5) $(x+2)(x+12) = 0$

6) $(5x+4)(x+7) = 0$

7) $(6x+1)(4x+5) = 0$

8) $(2x+7)(x+8) = 0$

9) $(x+6)(3x+15) = 0$

10) $(12x+2)(x+8) = 0$

11) $x^2 = 8x$

12) $x^2 - 16 = 0$

13) $3x^2 + 6 = 9x$

14) $-2x^2 - 8 = 10x$

15) $5x^2 + 40x = 45$

16) $x^2 + 10x = 24$

17) $x^2 + 6x = 16$

18) $x^2 + 9x = -18$

19) $x^2 + 13x = -36$

20) $x^2 + 3x - 15 = 5x$

21) $x^2 + 8x + 7 = -8$

22) $3x^2 - 11x = -9 + x$

23) $10x^2 + 3 = 27x - 15$

24) $7x^2 - 6x + 8 = 8$

25) $2x^2 - 12 = -3x + 2$

26) $10x^2 - 26x - 3 = -15$

27) $3x^2 + 21 = -16x + 5$

28) $x^2 + 15x - 10 = -66$

29) $3x^2 - 8x - 8 = 4 + x$

30) $2x^2 + 6x - 24 = 12$

31) $3x^2 - 33x + 54 = -18$

32) $-10x^2 - 15x - 9 = -9 - 27x^2$

WWW.MathNotion.Com

# Quadratic Formula and the Discriminant

✎ Find the value of the discriminant of each quadratic equation.

1) $3x(x - 8) = 0$

2) $2x^2 + 6x - 4 = 0$

3) $x^2 + 6x + 7 = 0$

4) $x^2 - x + 3 = 0$

5) $x^2 + 4x - 3 = 0$

6) $2x^2 + 6x - 10 = 0$

7) $3x^2 + 7x + 5 = 0$

8) $x^2 - 6x - 4 = 0$

9) $2x^2 + 8x + 3 = 0$

10) $x^2 + 7x - 5 = 0$

11) $5x^2 + 2x - 3 = 0$

12) $-3x^2 - 11x + 4 = 0$

13) $-6x^2 - 12x + 8 = 0$

14) $-x^2 - 9x - 12 = 0$

15) $7x^2 - 6x - 10 = 0$

16) $-4x^2 - 2x + 8 = 0$

17) $5x^2 + 8x - 2 = 0$

18) $6x^2 - 4x = 0$

19) $3x^2 - 5x + 2 = 0$

20) $4x^2 + 9x + 3 = 0$

✎ Find the discriminant of each quadratic equation then state the number of real and imaginary solutions.

21) $-4x^2 - 16 = 16x$

22) $20x^2 = 20x - 5$

23) $-11x^2 - 19x = 26$

24) $22x^2 - 4x + 1 = 18x^2$

25) $-11x^2 = -15x + 8$

26) $3x^2 + 6x + 9 = 6$

27) $13x^2 - 5x - 12 = -26$

28) $-8x^2 - 32x - 25 = 7$

# ACT Aspire Subject Test Mathematics Grade 8

## Graphing Quadratic Functions

✎ Sketch the graph of each function. Identify the vertex and axis of symmetry.

1) $y = (x + 3)^2 + 2$

2) $y = (x - 3)^2 - 2$

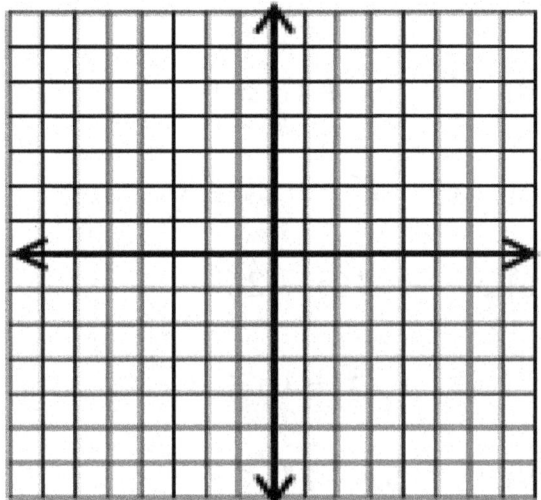

3) $y = 6 - (-x + 4)^2$

4) $y = -3x^2 - 6x + 9$

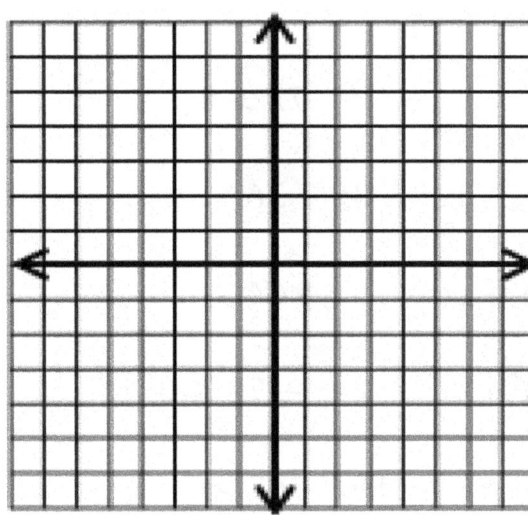

WWW.MathNotion.Com

# Quadratic Inequalities

✏️ **Solve each quadratic inequality.**

1) $x^2 - 25 < 0$

2) $-x^2 - 6x - 8 > 0$

3) $5x^2 + 15x + 30 < 0$

4) $x^2 + 8x + 16 > 0$

5) $2x^2 - 18x - 20 \geq 0$

6) $x^2 > -10x - 25$

7) $3x^2 + 2x + 16 \leq 0$

8) $x^2 - 5x - 14 \leq 0$

9) $x^2 - 6x - 7 \geq 0$

10) $2x^2 + 16x - 18 < 0$

11) $x^2 + 6x - 72 > 0$

12) $3x^2 - 3x - 36 > 0$

13) $x^2 - 15x + 64 \leq 0$

14) $2x^2 - 24x + 72 \leq 0$

15) $x^2 - 16x + 63 \geq 0$

16) $x^2 - 16x + 55 \geq 0$

17) $x^2 - 81 \leq 0$

18) $x^2 - 17x + 42 \geq 0$

19) $9x^2 + 14x + 36 \leq 0$

20) $4x^2 - 2x - 24 > 2x^2$

21) $5x^2 - 20x + 20 < 0$

22) $7x^2 - 6x \geq 6x^2 - 5$

23) $5x^2 - 15 > 4x^2 + 2x$

24) $3x^2 - 4x \geq 3x^2 - 9x + 15$

25) $8x^2 + 9x - 54 > 5x^2$

26) $10x^2 + 50x - 60 < 0$

27) $-x^2 + 15x - 57 \geq 0$

28) $-5x^2 + 25x + 30 \leq 0$

29) $5x^2 + 40x + 75 < 0$

30) $9x^2 + 20x + 180 \leq 0$

31) $3x^2 + 2x - 36 \geq -x$

32) $3x^2 + 9x + 9 \leq 6x^2 + 3x$

# ACT Aspire Subject Test Mathematics Grade 8

## Domain and Range of Radical Functions

✎ **Identify the domain and range of each function.**

1) $y = \sqrt{x+8} - 7$

2) $y = \sqrt[3]{3x-5} - 4$

3) $y = \sqrt{3x-9} + 3$

4) $y = \sqrt[3]{(4x+6)} - 2$

5) $y = 3\sqrt{4x+20} + 6$

6) $y = \sqrt[3]{(5x-2)} - 11$

7) $y = 4\sqrt{9x^2+8} + 3$

8) $y = \sqrt[3]{(7x^2-2)} - 6$

9) $y = 2\sqrt{2x^3+16} - 3$

10) $y = \sqrt[3]{(11x+4)} - 2x$

11) $y = 3\sqrt{-2(4x+8)} + 5$

12) $y = \sqrt[5]{(3x^2-12)} - 6$

13) $y = 3\sqrt{x-5} - 2$

14) $y = \sqrt[3]{6x+9} - 4$

✎ **Sketch the graph of each function.**

15) $y = -3\sqrt{x} + 5$

16) $y = 3\sqrt{x} - 6$

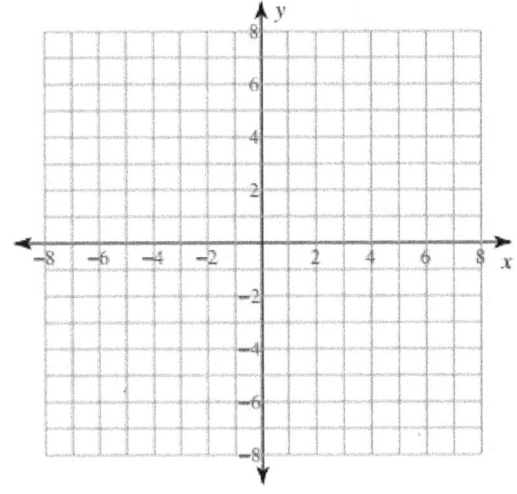

ACT Aspire Subject Test Mathematics Grade 8

## Solving Radical Equations

✎ Solve each equation. Remember to check for extraneous solutions.

1) $\sqrt{a} = 9$

2) $\sqrt{v} = 6$

3) $\sqrt{r} = 4$

4) $8 = 16\sqrt{x}$

5) $\sqrt{x+3} = 18$

6) $6 = \sqrt{x-7}$

7) $4 = \sqrt{r-3}$

8) $\sqrt{x-5} = 7$

9) $12 = \sqrt{x-4}$

10) $\sqrt{m+5} = 8$

11) $7\sqrt{5a} = 35$

12) $6\sqrt{2x} = 48$

13) $2 = \sqrt{6x-32}$

14) $\sqrt{304-4x} = 4$

15) $\sqrt{r+2} - 8 = 6$

16) $-21 = -7\sqrt{r+9}$

17) $60 = 6\sqrt{5v}$

18) $x = \sqrt{40-3x}$

19) $\sqrt{90-27a} = 3a$

20) $\sqrt{-8n+88} = 4$

21) $\sqrt{15r-5} = 4r-3$

22) $\sqrt{-64+32x} = 4x$

23) $\sqrt{4x+15} = \sqrt{2x+11}$

24) $\sqrt{12v} = \sqrt{15v-21}$

25) $\sqrt{9-x} = \sqrt{x-3}$

26) $\sqrt{6m+34} = \sqrt{8m+34}$

27) $\sqrt{7r+32} = \sqrt{-8-3r}$

28) $\sqrt{4k+10} = \sqrt{2-4k}$

29) $-20\sqrt{x-13} = -40$

30) $\sqrt{90-2x} = \sqrt{\dfrac{x}{4}}$

WWW.MathNotion.Com

# ACT Aspire Subject Test Mathematics Grade 8

## Answers of Worksheets

**Evaluating Function**

1) $h(x) = -8x + 3$
2) $k(a) = 2a - 14$
3) $d(t) = 11t$
4) $f(x) = \frac{5}{12}x - \frac{7}{12}$
5) $m(n) = 24n - 210$
6) $c(p) = p^2 - 5p + 10$
7) $-13$
8) $14$
9) $-3$
10) $22$
11) $9$
12) $26$
13) $2$
14) $7$
15) $-27$
16) $8.2$
17) $15$
18) $15$
19) $74$
20) $75$
21) $-1\frac{7}{8}$
22) $-3\frac{8}{9}$
23) $87$
24) $-28$
25) $14$
26) $-\frac{14b+9}{3b}$
27) $12a - 50$
28) $-2x + 18$
29) $2x^2 + 16$
30) $16x^4 - 20$

**Adding and Subtracting Functions**

1) $-2$
2) $1$
3) $-4$
4) $-3x^2 - 6x - 19$
5) $-43$
6) $15$
7) $-7\frac{2}{3}$
8) $4a^2 + 8a - 4$
9) $-18x^2 - 18x - 12$
10) $-2t^2 - 11t + 1$
11) $5x^4 - 5x^2 + 9$
12) $-81x^8 + 22$

**Multiplying and Dividing Functions**

1) $-55$
2) $-70$
3) $153$
4) $-1$
5) $4\frac{4}{7}$
6) $2$
7) $84$
8) $2$
9) $0$
10) $-62$
11) $6x^4 - 17x^3 + 5x^2 + 3x - 1$
12) $5x^8 - 2x^6 - 10x^4 + 4x^2$

**Composition of Functions**

1) $-13$
2) $-1$
3) $26$
4) $-10$
5) $-17$
6) $10$
7) $\frac{25}{8}$
8) $\frac{1}{8}$
9) $8$
10) $\frac{5}{8}$
11) $\frac{49}{8}$
12) $-\frac{1}{2}(x^2 - \frac{3}{2})$
13) $1$
14) $-2$
15) $2$
16) $6$
17) $-1$
18) $-2x$

WWW.MathNotion.Com

# ACT Aspire Subject Test Mathematics Grade 8

19) 3   21) 2   23) −7
20) −9   22) $\sqrt{13}$   24) −18

## Quadratic Equations

1) $x^2 + 2x - 24$
2) $x^2 + 12x + 35$
3) $x^2 + 2x - 48$
4) $x^2 - 7x - 18$
5) $x^2 - 15x + 56$
6) $3x^2 - 7x - 6$
7) $4x^2 + 5x - 6$
8) $4x^2 - x - 5$
9) $7x^2 - 41x - 6$
10) $15x^2 - 12x - 3$

11) $(x - 4)(x + 2)$
12) $(x + 5)(x + 3)$
13) $(x - 6)(x + 4)$
14) $(x - 3)(x - 7)$
15) $(x + 3)(x + 7)$
16) $(4x + 5)(x + 1)$
17) $(5x - 2)(x + 3)$
18) $(5x - 3)(x + 4)$
19) $(2x + 5)(x + 1)$
20) $3(x - 2)(3x - 1)$

21) $x = -6, x = 3$
22) $x = -1, x = -8$
23) $x = -2, x = -5$
24) $x = 1, x = -2$
25) $x = 3, x = -4$
26) $x = -3, x = -8$
27) $x = -4, x = -\frac{1}{2}$
28) $x = 4, x = -7$
29) $x = 3, x = -4$
30) $x = -3, x = 2$

## Solving quadratic equations

1) $\{-9, 1\}$
2) $\{-6, -7\}$
3) $\{8, -3\}$
4) $\{6, 4\}$
5) $\{-2, -12\}$
6) $\{-\frac{4}{5}, -7\}$
7) $\{-\frac{5}{4}, -\frac{1}{6}\}$
8) $\{-\frac{7}{2}, -8\}$

9) $\{-6, -5\}$
10) $\{-\frac{1}{6}, -8\}$
11) $\{8, 0\}$
12) $\{4, -4\}$
13) $\{2, 1\}$
14) $\{-4, -1\}$
15) $\{1, -9\}$
16) $\{2, -12\}$

17) $\{2, -8\}$
18) $\{-3, -6\}$
19) $\{-4, -9\}$
20) $\{5, -3\}$
21) $\{-5, -3\}$
22) $\{1, 3\}$
23) $\{\frac{6}{5}, \frac{3}{2}\}$
24) $\{\frac{6}{7}, 0\}$

25) $\{-\frac{7}{2}, 2\}$
26) $\{\frac{3}{5}, 2\}$
27) $\{-\frac{4}{3}, -4\}$
28) $\{-8, -7\}$
29) $\{4, -1\}$
30) $\{3, -6\}$
31) $\{3, 8\}$
32) $\{\frac{15}{17}, 0\}$

## Quadratic formula and the discriminant

1) 576
2) 68
3) 8
4) −11
5) 28

6) 116
7) −11
8) 52
9) 40
10) 69

11) 64
12) 169
13) 336
14) 33
15) 316

16) 132
17) 104
18) 16
19) 1
20) 33

21) 0, one real solution   22) 0, one real solution   23) −783, no solution

WWW.MathNotion.Com

# ACT Aspire Subject Test Mathematics Grade 8

24) 0, one real solution     26) 0, one real solution     28) 0, one real solution

25) $-127$, no solution     27) $-703$, no solution

**Graphing quadratic functions**

1) $(-3, 2), x = -3$

2) $(3, -2), x = 3$

3) $(4, 6), x = 4$

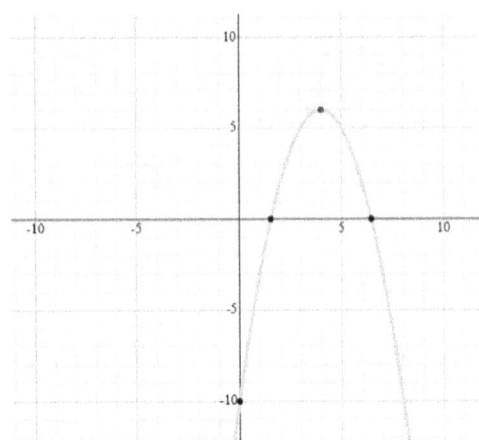

4) $(-1, 12), x = -1$

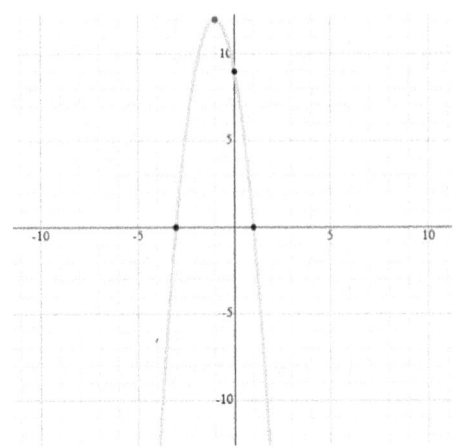

**Quadratic inequalities**

1) $-5 < x < 5$     7) no solution     13) no solution

2) $-4 < x < -2$     8) $-2 \leq x \leq 7$     14) $x = 6$

3) no solution     9) $x \leq -1 \text{ or } x \geq 7$     15) $x \leq 7 \text{ or } x \geq 9$

4) $x < -4 \text{ or } x > -4$     10) $-9 < x < 1$     16) $x \leq 5 \text{ or } x \geq 11$

5) $x \leq -1 \text{ or } x \geq 10$     11) $x < -12 \text{ or } x > 6$     17) $-9 \leq x \leq 9$

6) $x < -5 \text{ or } x > -5$     12) $-3 < x < 4$     18) $x \leq 3 \text{ or } x \geq 14$

# ACT Aspire Subject Test Mathematics Grade 8

19) no solution
20) $x < -3$ or $x > 4$
21) no solution
22) $x \leq 1$ or $x \geq 5$
23) $x < -3$ or $x > 5$
24) $x \geq 3$
25) $x < -6$ or $x > 3$
26) $-6 < x < 1$
27) no solution
28) $x \leq -1$ or $x \geq 6$
29) $-5 < x < -3$
30) no solution
31) $x \leq -4$ or $x \geq 3$
32) $x \leq -1$ or $x \geq 3$

**Domain and range of radical functions**

1) domain: $x \geq -8$
   range: $y \geq -7$
2) domain: {all real numbers}
   range: {all real numbers}
3) domain: $x \geq 3$
   range: $y \geq 3$
4) domain: {all real numbers}
   range: {all real numbers}
5) domain: $x \geq -5$
   range: $y \geq 6$
6) domain: {all real numbers}
   range: {all real numbers}
7) domain: {all real numbers}
   range: $y \geq 8\sqrt{2} + 3$
8) domain: {all real numbers}
   range: {all real numbers}
9) domain: $x \geq -2$
   range: $y \geq -3$
10) domain: {all real numbers}
    range: {all real numbers}
11) domain: $x \leq -2$
    range: $y \geq 5$
12) domain: {all real numbers}
    range: {all real numbers}
13) domain: $x \geq 5$
    range: $y \geq -2$
14) domain: {all real numbers}
    range: {all real numbers}

15)

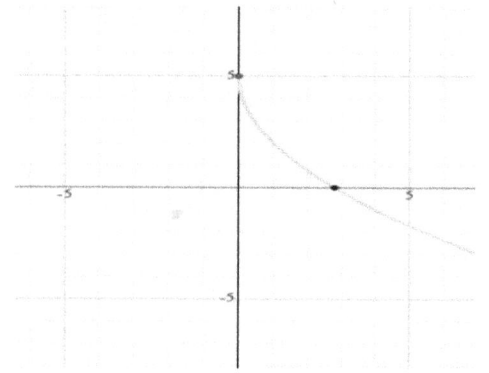

16)

# ACT Aspire Subject Test Mathematics Grade 8

**Solving radical equations**

1) {81}
2) {36}
3) {16}
4) {$\frac{1}{4}$}
5) {321}
6) {43}
7) {19}
8) {54}
9) {148}
10) {59}
11) {5}
12) {32}
13) {6}
14) {72}
15) {194}
16) {0}
17) {20}
18) {5}
19) {2}
20) {9}
21) {2}
22) no solution
23) {−2}
24) {7}
25) {6}
26) {0}
27) {−4}
28) {−1}
29) {17}
30) {40}

ACT Aspire Subject Test Mathematics Grade 8

# Chapter 10 :

# Geometry and Solid Figures

**Topics that you will practice in this chapter:**

- ✓ Angles
- ✓ Pythagorean Relationship
- ✓ Triangles
- ✓ Polygons
- ✓ Trapezoids
- ✓ Circles
- ✓ Cubes
- ✓ Rectangular Prism
- ✓ Cylinder
- ✓ Pyramids and Cone

*Mathematics is, as it were, a sensuous logic, and relates to philosophy as do the arts, music, and plastic art to poetry.* — *K. Shegel*

# ACT Aspire Subject Test Mathematics Grade 8

## Angles

✎ **What is the value of $x$ in the following figures?**

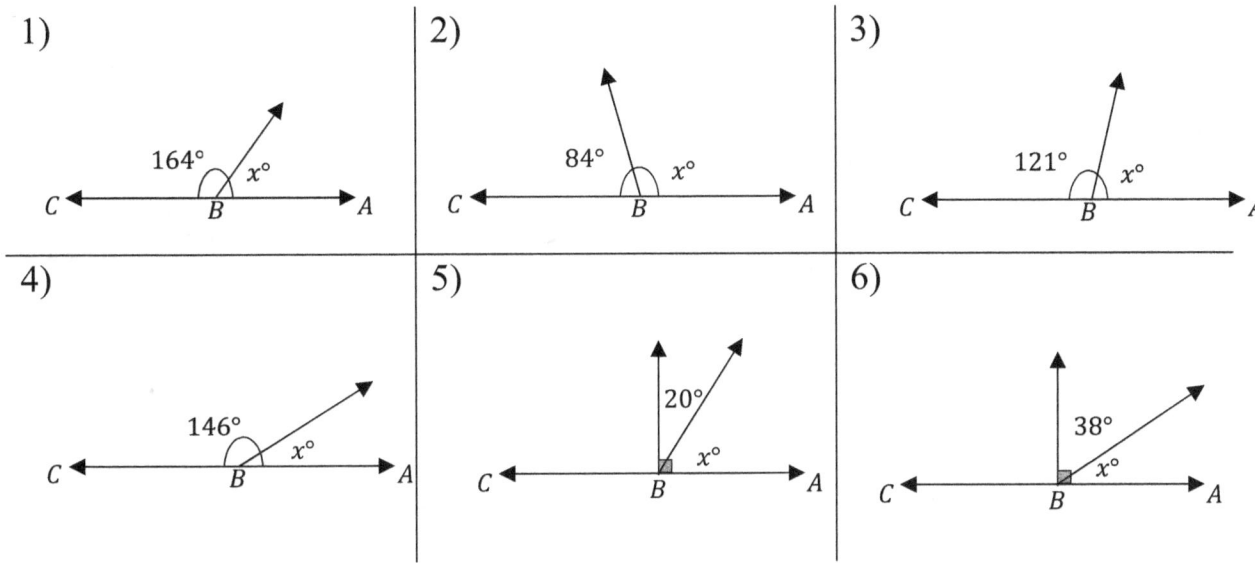

✎ **Calculate.**

7) Two supplement angles have equal measures. What is the measure of each angle? _____

8) The measure of an angle is seven fifth the measure of its supplement. What is the measure of the angle? _____

9) Two angles are complementary and the measure of one angle is 24 less than the other. What is the measure of the smaller angle? _____

10) Two angles are complementary. The measure of one angle is one fifth the measure of the other. What is the measure of the bigger angle? _____

11) Two supplementary angles are given. The measure of one angle is 40° less than the measure of the other. What does the smaller angle measure? _____

WWW.MathNotion.Com

# ACT Aspire Subject Test Mathematics Grade 8

## Pythagorean Relationship

✍ **Do the following lengths form a right triangle?**

1)

2)

3)

4)

5)

6)

7)

8)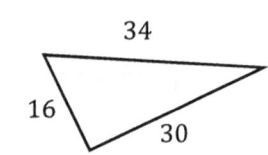

✍ **Find the missing side?**

9)

10)

11)

12)

13)

14)

15)

16)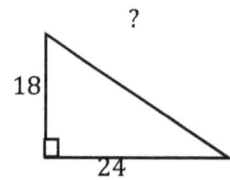

WWW.MathNotion.Com

# ACT Aspire Subject Test Mathematics Grade 8

## Triangles

✎ Find the measure of the unknown angle in each triangle.

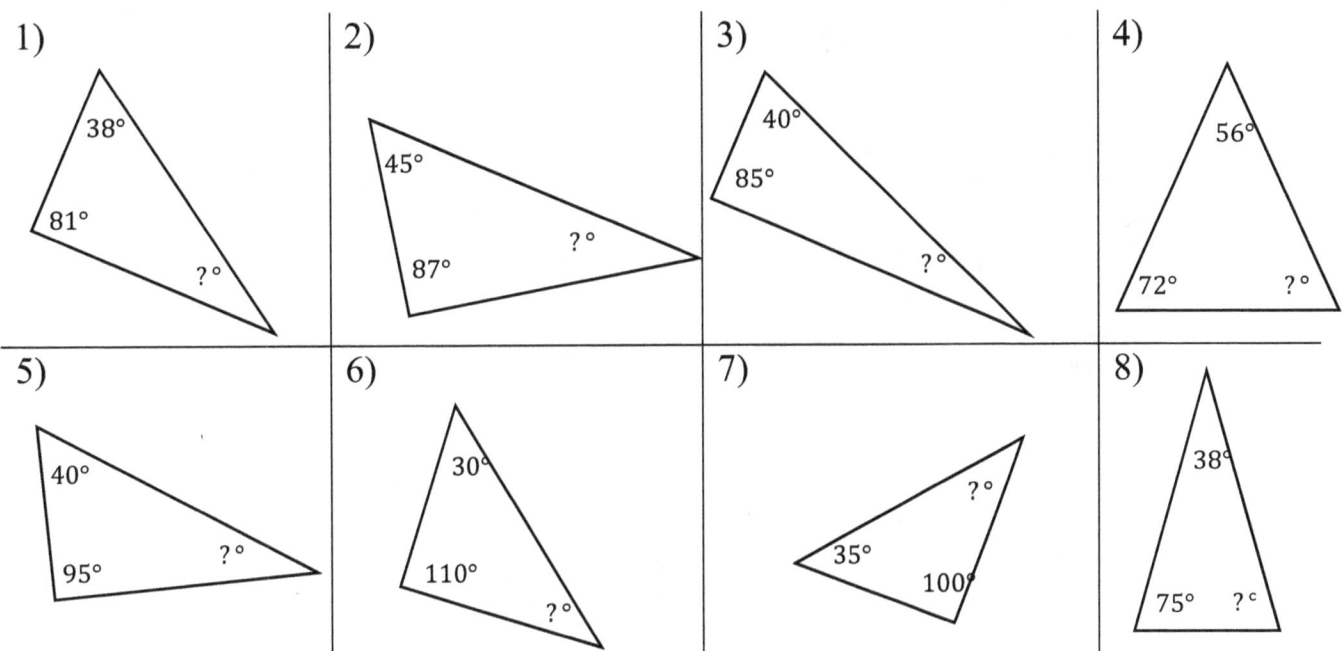

✎ Find area of each triangle.

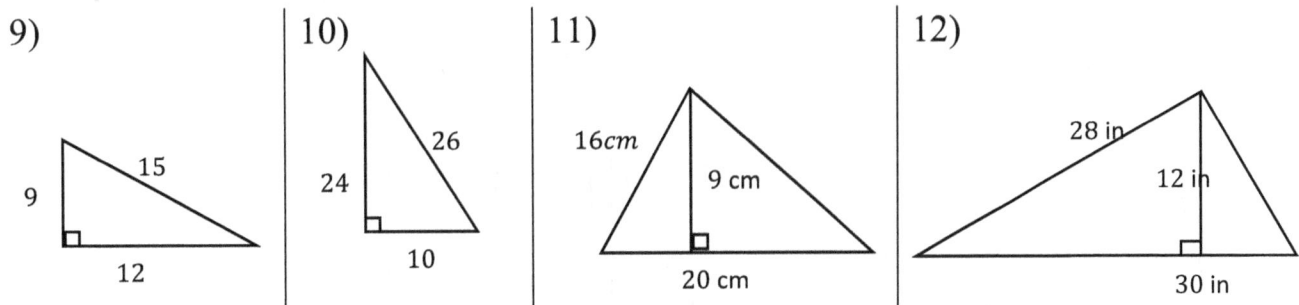

WWW.MathNotion.Com

# ACT Aspire Subject Test Mathematics Grade 8

## Polygons

✍ Find the perimeter of each shape.

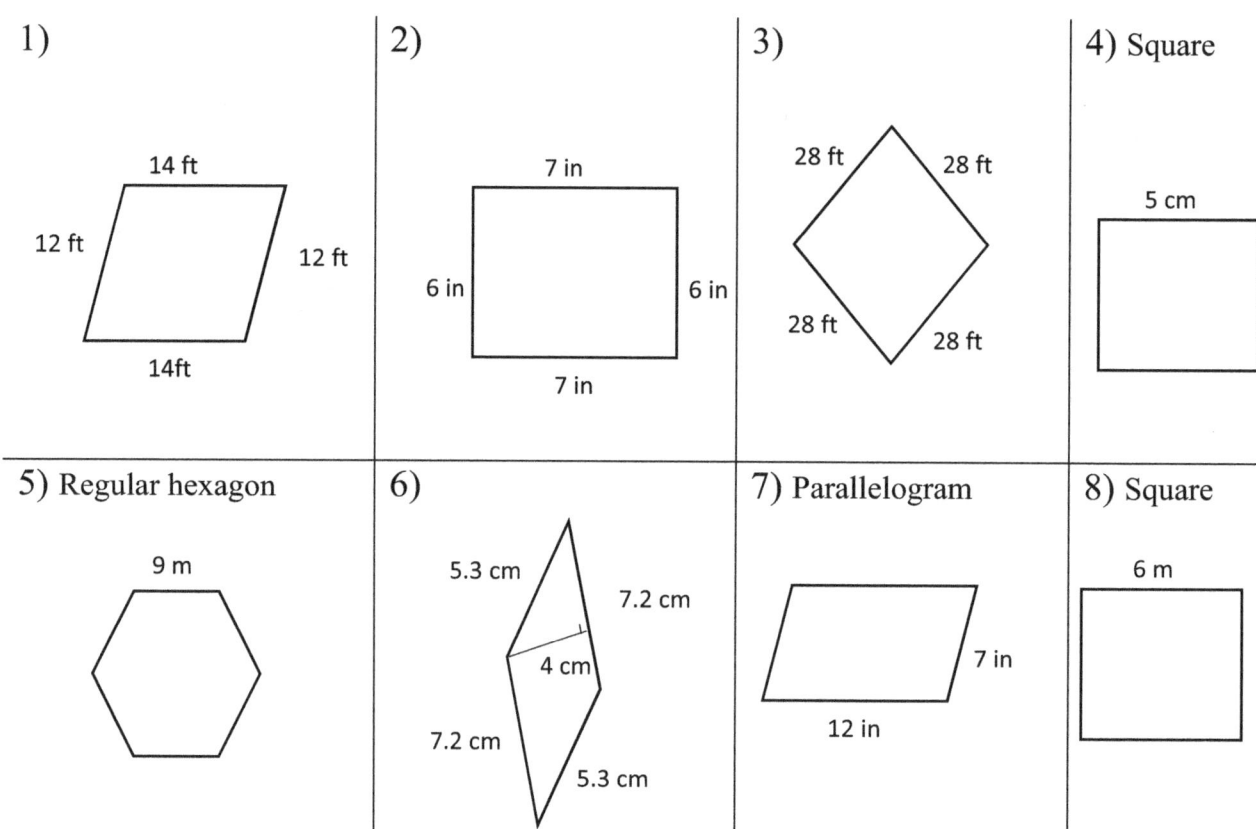

✍ Find the area of each shape.

9) Parallelogram   10) Rectangle   11) Rectangle   12) Square

# Trapezoids

🖋 **Find the area of each trapezoid.**

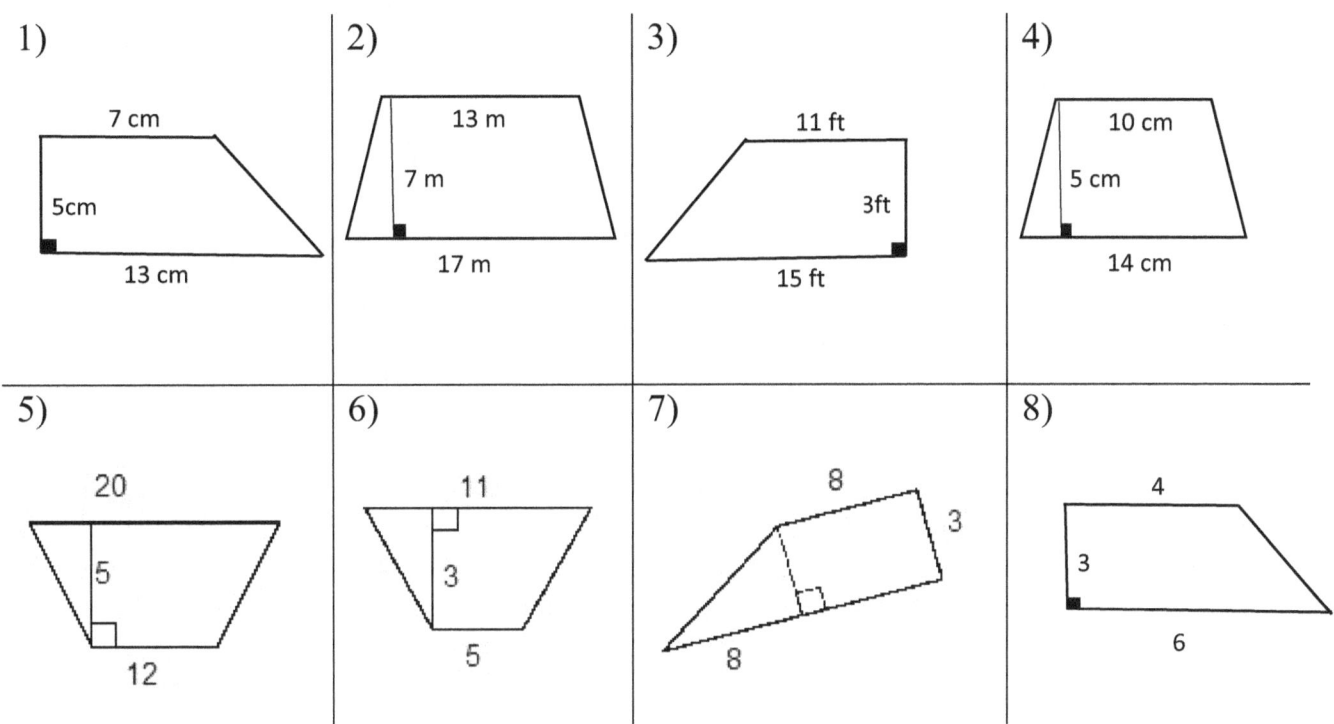

🖋 **Calculate.**

1) A trapezoid has an area of 45 cm² and its height is 5 cm and one base is 5 cm. What is the other base length? _____

2) If a trapezoid has an area of 99 ft² and the lengths of the bases are 8 ft and 10 ft, find the height? _____

3) If a trapezoid has an area of 126 m² and its height is 14 m and one base is 6 m, find the other base length? _____

4) The area of a trapezoid is 440 ft² and its height is 22 ft. If one base of the trapezoid is 15 ft, what is the other base length?
_____

# ACT Aspire Subject Test Mathematics Grade 8

## Circles

✏️ **Find the area of each circle.** ($\pi = 3.14$)

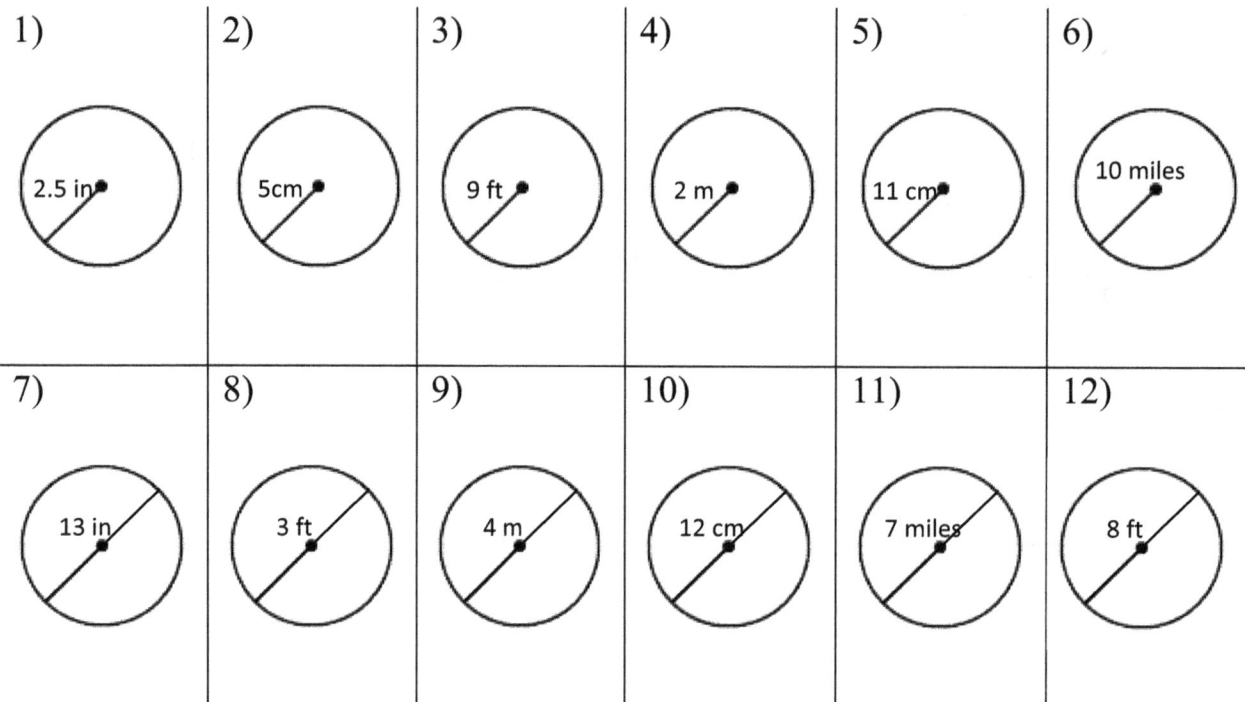

1) 2.5 in
2) 5 cm
3) 9 ft
4) 2 m
5) 11 cm
6) 10 miles
7) 13 in
8) 3 ft
9) 4 m
10) 12 cm
11) 7 miles
12) 8 ft

✏️ **Complete the table below.** ($\pi = 3.14$)

| Circle No. | Radius | Diameter | Circumference | Area |
|---|---|---|---|---|
| 1 | 1 in | 2 in | 6.28 in | 3.14 $in^2$ |
| 2 |  | 10 m |  |  |
| 3 |  |  |  | 28.26 $ft^2$ |
| 4 |  |  | 47.1 mi |  |
| 5 |  | 11 km |  |  |
| 6 | 7 cm |  |  |  |
| 7 |  | 12 ft |  |  |
| 8 |  |  |  | 314 $m^2$ |
| 9 |  |  | 56.52 in |  |
| 10 | 4.5 ft |  |  |  |

WWW.MathNotion.Com

# ACT Aspire Subject Test Mathematics Grade 8

## Cubes

✎ Find the volume of each cube.

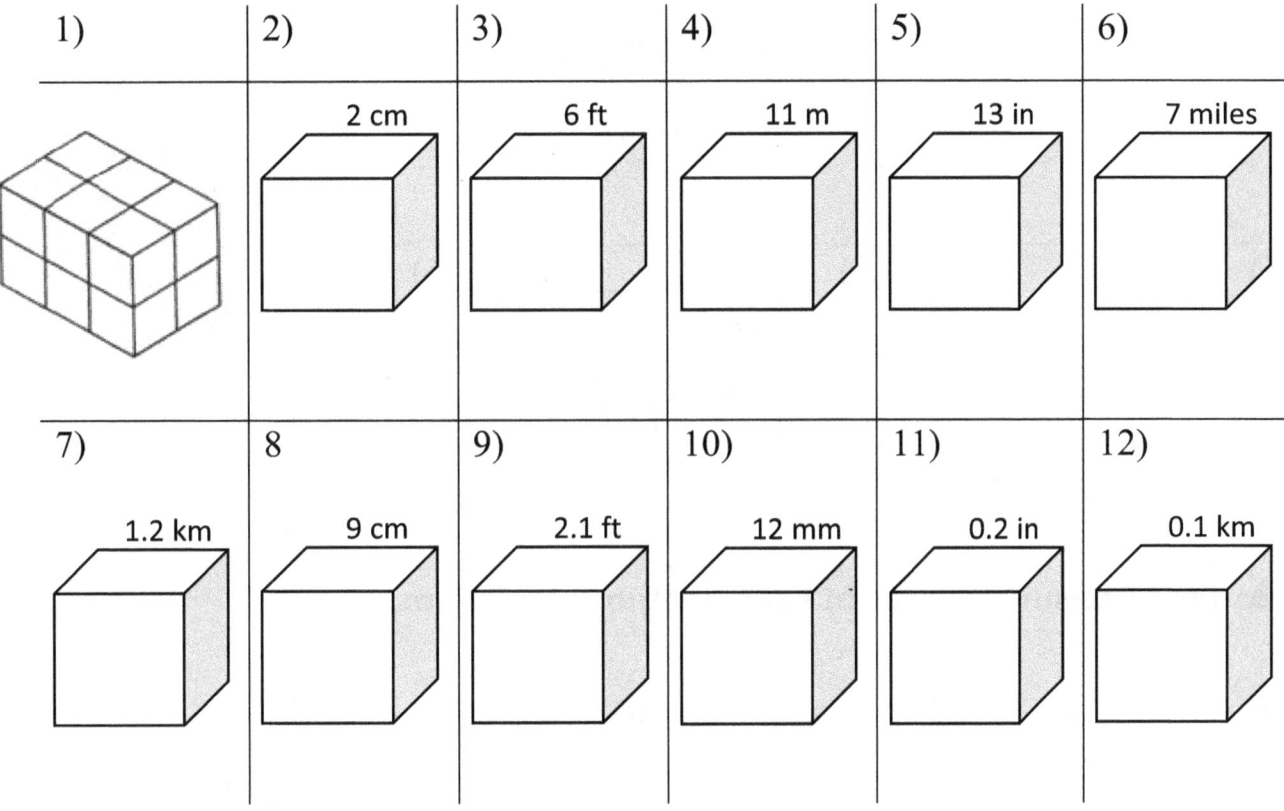

✎ Find the surface area of each cube.

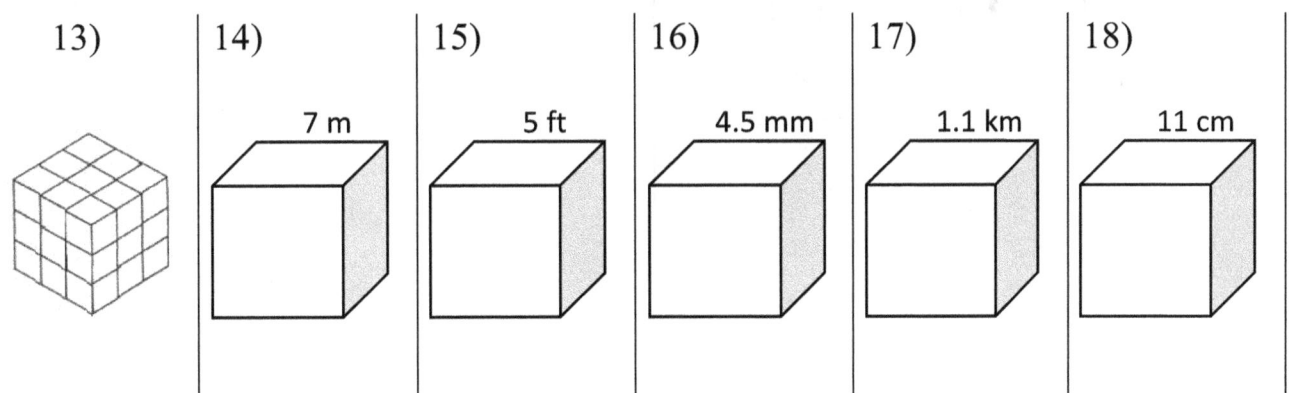

## Rectangular Prism

✎ Find the volume of each Rectangular Prism.

# ACT Aspire Subject Test Mathematics Grade 8

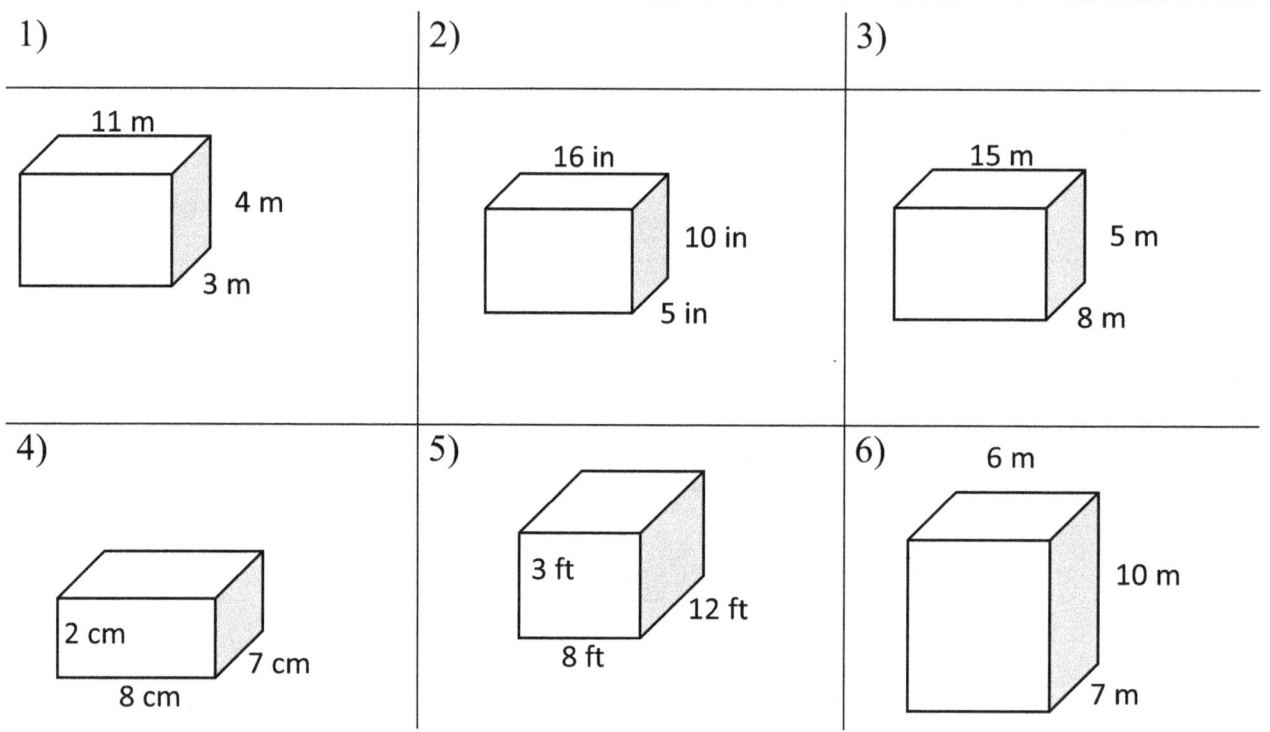

🖎 **Find the surface area of each Rectangular Prism.**

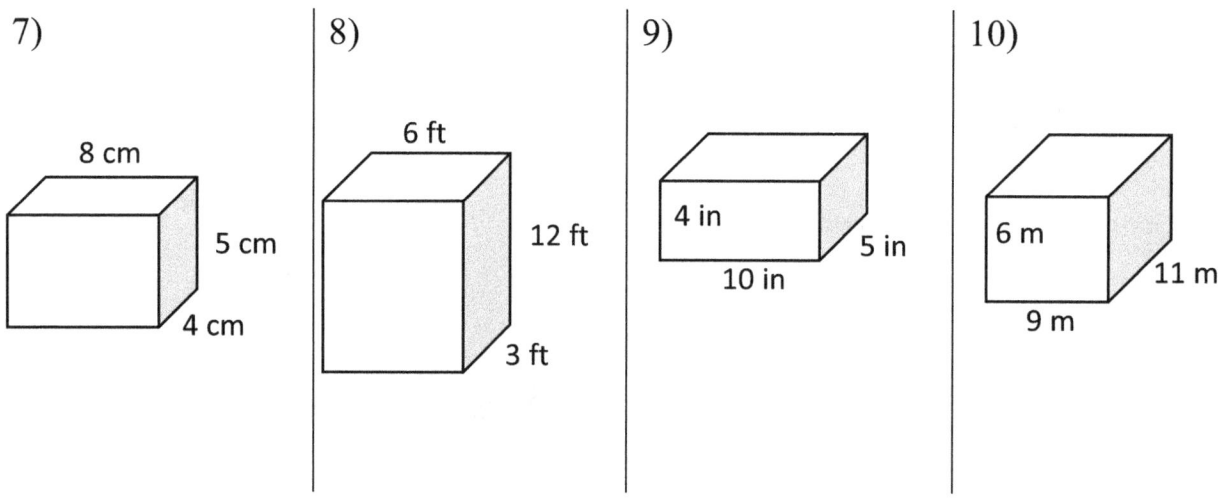

# ACT Aspire Subject Test Mathematics Grade 8

## Cylinder

✎ **Find the volume of each Cylinder. Round your answer to the nearest tenth.** ($\pi = 3.14$)

1)

2)

3)

4)

5)

6)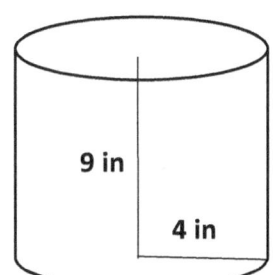

✎ **Find the surface area of each Cylinder.** ($\pi = 3.14$)

7)

8)

9)

10)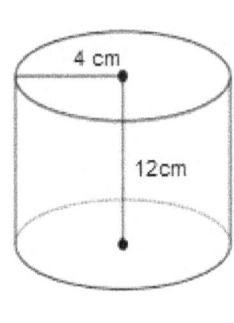

WWW.MathNotion.Com

# ACT Aspire Subject Test Mathematics Grade 8

## Pyramids and Cone

🍃 **Find the volume of each Pyramid and Cone.** ($\pi = 3.14$)

1)

2)

3)

4)

5)

6)

🍃 **Find the surface area of each Pyramid and Cone.** ($\pi = 3.14$)

7)

8)

9)

10)

WWW.MathNotion.Com

# ACT Aspire Subject Test Mathematics Grade 8

## Answers of Worksheets

**Angles**

1) 16°    4) 34°    7) 90°    10) 75°
2) 96°    5) 70°    8) 75°    11) 70°
3) 59°    6) 52°    9) 33°

**Pythagorean Relationship**

1) No     5) Yes    9) 13     13) 15
2) Yes    6) No     10) 20    14) 30
3) No     7) Yes    11) 17    15) 36
4) Yes    8) Yes    12) 10    16) 12

**Triangles**

1) 60°    5) 45°    9) 54 square unites
2) 48°    6) 40°    10) 120 square unites
3) 55°    7) 45°    11) 90 square unites
4) 52°    8) 67°    12) 180 square unites

**Polygons**

1) 52 ft    5) 54 m     9) 30 $m^2$
2) 26 in    6) 25 cm    10) 300 $in^2$
3) 112 ft   7) 38 in    11) 160 $km^2$
4) 20 cm    8) 24 m     12) 49 $in^2$

**Trapezoids**

1) 50 $cm^2$    4) 60 $cm^2$    7) 36
2) 105 $m^2$    5) 80           8) 15
3) 39 $ft^2$    6) 24

**Calculate**

1) 13 cm    2) 11 ft    3) 12 m    4) 25 ft

**Circles**

1) 19.63 $in^2$    5) 379.94 $cm^2$    9) 12.56 $m^2$
2) 78.5 $cm^2$     6) 314 $miles^2$    10) 113.04 $cm^2$
3) 254.34 $ft^2$   7) 132.67 $in^2$    11) 38.47 $miles^2$
4) 12.56 $m^2$     8) 7.07 $ft^2$      12) 50.24 $ft^2$

WWW.MathNotion.Com

# ACT Aspire Subject Test Mathematics Grade 8

| Circle No. | Radius | Diameter | Circumference | Area |
|---|---|---|---|---|
| 1 | 1 in | 2 in | 6.28 in | 3.14 $in^2$ |
| 2 | 5 m | 10 m | 31.4 m | 78.5 $m^2$ |
| 3 | 3 ft | 6 ft | 18.84 ft | 28.26 $ft^2$ |
| 4 | 7.5 miles | 15 mi | 47.1 mi | 176.63 $mi^2$ |
| 5 | 5.5 km | 11 km | 34.54 km | 94.99 $km^2$ |
| 6 | 7 cm | 14 cm | 43.96 cm | 153.86 $cm^2$ |
| 7 | 6 ft | 12 ft | 37.68 feet | 113.04 $ft^2$ |
| 8 | 10 m | 20 m | 62.8 m | 314 $m^2$ |
| 9 | 9 in | 18 in | 56.52 in | 254.34 $in^2$ |
| 10 | 4.5 ft | 9 ft | 28.26 ft | 63.585 $ft^2$ |

**Cubes**

1) 12
2) 8 $cm^3$
3) 216 $ft^3$
4) 1,331 $m^3$
5) 2,197 $in^3$
6) 343 $miles^3$
7) 1.728 $km^3$
8) 729 $cm^3$
9) 9.261 $ft^3$
10) 1,728 $mm^3$
11) 0.008 $in^3$
12) 0.001 $km^3$
13) 27
14) 294 $m^2$
15) 150 $ft^2$
16) 121.5 $mm^2$
17) 7.26 $km^2$
18) 726 $cm^2$

**Rectangular Prism**

1) 132 $m^3$
2) 800 $in^3$
3) 600 $m^3$
4) 112 $cm^3$
5) 288 $ft^3$
6) 420 $m^3$
7) 184 $cm^2$
8) 252 $ft^2$
9) 220 $in^2$
10) 438 $m^2$

**Cylinder**

1) 1,004.8 $m^3$
2) 214.6 $cm^3$
3) 9,495.4 $cm^3$
4) 1.1 $m^3$
5) 588.8 $m^3$
6) 452.2 $in^3$
7) 188.4 $m^2$
8) 602.9 $cm^2$
9) 37.7 $cm^2$
10) 401.9 $m^2$

**Pyramids and Cone**

1) 1,600 $yd^3$
2) 1,050 $yd^3$
3) 1,617 $in^3$
4) 392.5 $m^3$
5) 3,014.4 $m^3$
6) 366.33 $cm^3$
7) 1,440 $yd^2$
8) 1,536 $m^2$
9) 678.24 $in^2$
10) 1,205.76 $cm^2$

ACT Aspire Subject Test Mathematics Grade 8

# Chapter 11:
# Statistics and Probability

**Topics that you will practice in this chapter:**

- ✓ Mean and Median
- ✓ Mode and Range
- ✓ Histograms
- ✓ Stem–and–Leaf Plot
- ✓ Pie Graph
- ✓ Probability Problems
- ✓ Factorials
- ✓ Combinations and Permutation

*Mathematics is no more computation than typing is literature.*
*– John Allen Paulos*

# ACT Aspire Subject Test Mathematics Grade 8

## Mean and Median

✍ **Find Mean and Median of the Given Data.**

1) 8, 7, 14, 4, 8

2) 14, 8, 25, 19, 16, 33, 11

3) 23, 18, 15, 12, 17

4) 34, 14, 10, 15, 6, 11

5) 10, 19, 6, 8, 32, 20, 17

6) 17, 26, 39, 69, 20, 6

7) 40, 38, 18, 11, 9, 2, 7, 32, 41

8) 24, 21, 31, 12, 33, 32, 22

9) 16, 14, 20, 41, 15, 20, 38, 4

10) 20, 20, 30, 18, 6, 28, 12, 46

11) 12, 7, 10, 11, 16, 22

12) 10, 29, 27, 12, 2, 15, 10, 3

✍ **Calculate.**

13) In a javelin throw competition, five athletics score 56, 34, 62, 23 and 19 meters. What are their Mean and Median? _____

14) Eva went to shop and bought 8 apples, 14 peaches, 6 bananas, 4 pineapples and 12 melons. What are the Mean and Median of her purchase? _____

15) Bob has 17 black pen, 19 red pen, 14 green pens, 20 blue pens and 5 boxes of yellow pens. If the Mean and Median are 19 respectively, what is the number of yellow pens in each box? _____

# ACT Aspire Subject Test Mathematics Grade 8

## Mode and Range

✎ **Find Mode and Rage of the Given Data.**

1) 4, 3, 7, 3, 3, 4
   Mode: _____   Range: _____

2) 18, 18, 24, 26, 18, 8, 14, 22
   Mode: _____   Range: _____

3) 8, 8, 8, 16, 19, 22, 20, 9, 13
   Mode: _____   Range: _____

4) 24, 24, 14, 28, 20, 18, 20, 24
   Mode: _____   Range: _____

5) 6, 21, 27, 24, 27, 27
   Mode: _____   Range: _____

6) 21, 8, 8, 7, 8, 12, 10, 22, 18, 13
   Mode: _____   Range: _____

7) 7, 4, 4, 6, 13, 13, 13, 0, 2, 2
   Mode: _____   Range: _____

8) 5, 8, 5, 14, 12, 14, 3, 5, 18
   Mode: _____   Range: _____

9) 7, 7, 7, 12, 7, 3, 8, 16, 3, 17
   Mode: _____   Range: _____

10) 15, 15, 19, 16, 4, 16, 10, 15
    Mode: _____   Range: _____

11) 6, 6, 5, 6, 42, 13, 19, 2
    Mode: _____   Range: _____

12) 8, 8, 9, 8, 9, 4, 34, 22
    Mode: _____   Range: _____

✎ **Calculate.**

13) A stationery sold 12 pencils, 56 red pens, 24 blue pens, 20 notebooks, 12 erasers, 21 rulers and 11 color pencils. What are the Mode and Range for the stationery sells?

   Mode: _____   Range: _____

14) In an English test, eight students score 10, 15, 15, 18 18, 16, 15 and 15. What are their Mode and Range? _____

15) What is the range of the first 6 even numbers greater than 8? _____

WWW.MathNotion.Com

# ACT Aspire Subject Test Mathematics Grade 8

## Times Series

✎ **Use the following Graph to complete the table.**

| Day | Distance (km) |
|-----|---------------|
| 1   |               |
| 2   |               |
|     |               |
|     |               |
|     |               |
|     |               |

The following table shows the number of births in the US from 2007 to 2012 (in millions).

| Year | Number of births (in millions) |
|------|-------------------------------|
| 2007 | 4.15 |
| 2008 | 3.70 |
| 2009 | 3.45 |
| 2010 | 3.20 |
| 2011 | 1.75 |
| 2012 | 2.98 |

Draw a Time Series for the table.

# ACT Aspire Subject Test Mathematics Grade 8

## Stem–and–Leaf Plot

✏️ **Make stem ad leaf plots for the given data.**

1) 24, 26, 29, 20, 53, 27, 51, 55, 36, 21, 37, 30

   Stem | Leaf plot

2) 11, 59, 66, 14, 18, 19, 59, 65, 69, 61, 68, 65

   Stem | Leaf plot

3) 121, 55, 66, 54, 112, 128, 63, 125, 59, 123, 68, 119

   Stem | Leaf plot

4) 51, 32, 100, 56, 84, 36, 107, 56, 85, 39, 56, 106, 89

   Stem | Leaf plot

5) 33, 89, 19, 87, 81, 16, 11, 30, 86, 35, 17, 35, 13

   Stem | Leaf plot

6) 60, 92, 22, 25, 67, 93, 95, 62, 21, 64, 98, 29

   Stem | Leaf plot

# Pie Graph

The circle graph below shows all Robert's expenses for last month. Robert spent $140 on his hobbies last month.

Answer following questions based on the Pie graph.

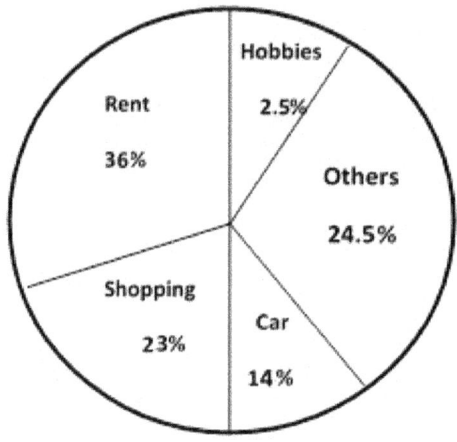

1) How much was Robert's total expenses last month? _____

2) How much did Robert spend on his car last month? _____

3) How much did Robert spend for shopping last month? _____

4) How much did Robert spend on his rent last month? _____

5) What fraction is Robert's expenses for his rent and car out of his total expenses last month? _____

ACT Aspire Subject Test Mathematics Grade 8

# Probability Problems

✎ **Calculate.**

1) A number is chosen at random from 1 to 10. Find the probability of selecting number 6 or smaller numbers. _____

2) Bag A contains 18 red marbles and 6 green marbles. Bag B contains 16 black marbles and 8 orange marbles. What is the probability of selecting a green marble at random from bag A? What is the probability of selecting a black marble at random from Bag B? _____

3) A number is chosen at random from 1 to 20. What is the probability of selecting multiples of 4? _____

4) A card is chosen from a well-shuffled deck of 52 cards. What is the probability that the card will be a queen? _____

5) A number is chosen at random from 1 to 15. What is the probability of selecting a multiple of 3 or 5? _____

A spinner numbered 1–8, is spun once. What is the probability of spinning …?

6) an Odd number? _____   7) a multiple of 2? _____

8) a multiple of 5? _____   9) number 10? _____

# ACT Aspire Subject Test Mathematics Grade 8

## Factorials

✎ **Determine the value for each expression.**

1) $4! + 0! =$

2) $2! + 5! =$

3) $(2!)^2 =$

4) $5! - 3! =$

5) $6! - 3! + 10 =$

6) $3! \times 4 - 15 =$

7) $(2! + 3!)^2 =$

8) $(4! - 3!)^2 =$

9) $(3!\,0!)^2 - 10 =$

10) $\dfrac{10!}{8!} =$

11) $\dfrac{6!}{4!} =$

12) $\dfrac{6!}{5!} =$

13) $\dfrac{15!}{13!} =$

14) $\dfrac{n!}{(n-3)!} =$

15) $\dfrac{(n+2)!}{n!} =$

16) $\dfrac{(2+2!)^3}{2!} =$

17) $\dfrac{5(n+2)!}{(n+1)!} =$

18) $\dfrac{22!}{20!\,4!} =$

19) $\dfrac{13!}{11!\,3!} =$

20) $\dfrac{9 \times 210!}{3(7 \times 30)!} =$

21) $\dfrac{32!}{31!\,2!} =$

22) $\dfrac{11!\,12!}{10!\,13!} =$

23) $\dfrac{16!\,15!}{14!\,14!} =$

24) $\dfrac{(5 \times 3)!}{0!\,14!} =$

25) $\dfrac{4!(5n-2)!}{(5n)!} =$

26) $\dfrac{4n(4n+7)!}{(4n+8)!} =$

27) $\dfrac{(n-2)!(n+1)}{(n+2)!} =$

WWW.MathNotion.Com

# ACT Aspire Subject Test Mathematics Grade 8

## Combinations and Permutations

✏ **Calculate the value of each.**

1) $6! =$ _____

2) $2! \times 5! =$ _____

3) $3 \times 4! =$ _____

4) $5! + 3! =$ _____

5) $7! =$ _____

6) $4! =$ _____

7) $3! + 3! =$ _____

8) $7! - 5! =$ _____

✏ **Find the answer for each word problems.**

9) Susan is baking cookies. She uses sugar, butter, Vanilla, eggs and flour. How many different orders of ingredients can she try? _____

10) Albert is planning for his vacation. He wants to go to museum, watch a movie, go to the beach, play the game and play football. How many ways of ordering are there for him? _____

11) How many 4-digit numbers can be named using the digits 3, 4, 5, and 6 without repetition? _____

12) In how many ways can 5 boys be arranged in a straight line? _____

13) In how many ways can 6 athletes be arranged in a straight line? _____

14) A professor is going to arrange her 7 students in a straight line. In how many ways can she do this? _____

15) How many code symbols can be formed with the letters for the word GAMES? _____

16) In how many ways a team of 7 basketball players can choose a captain and co-captain? _____

WWW.MathNotion.Com

# ACT Aspire Subject Test Mathematics Grade 8

## Answers of Worksheets

**Mean and Median**

1) Mean: 8.2, Median: 8
2) Mean: 18, Median: 16
3) Mean: 17, Median: 17
4) Mean: 15, Median: 12.5
5) Mean: 16, Median: 17
6) Mean: 29.5, Median: 23
7) Mean: 22, Median: 18
8) Mean: 25, Median: 24
9) Mean: 21, Median: 18
10) Mean: 22.5, Median: 20
11) Mean: 13, Median: 11.5
12) Mean: 13.5, Median: 11
13) Mean: 38.8, Median: 34
14) Mean: 8.8, Median: 8
15) 5

**Mode and Range**

1) Mode: 3, Range: 4
2) Mode: 18, Range: 18
3) Mode: 8, Range: 14
4) Mode: 24, Range: 14
5) Mode: 27, Range: 21
6) Mode: 8, Range: 15
7) Mode: 13, Range: 13
8) Mode: 5, Range: 15
9) Mode: 7, Range: 14
10) Mode: 15, Range: 15
11) Mode: 6, Range: 40
12) Mode: 8, Range: 30
13) Mode: 12, Range: 45
14) Mode: 15, Range: 8
15) 10

**Time series**

| Day | Distance (km) |
|---|---|
| 1 | 335 |
| 2 | 496 |
| 3 | 270 |
| 4 | 610 |
| 5 | 320 |
| 6 | 400 |

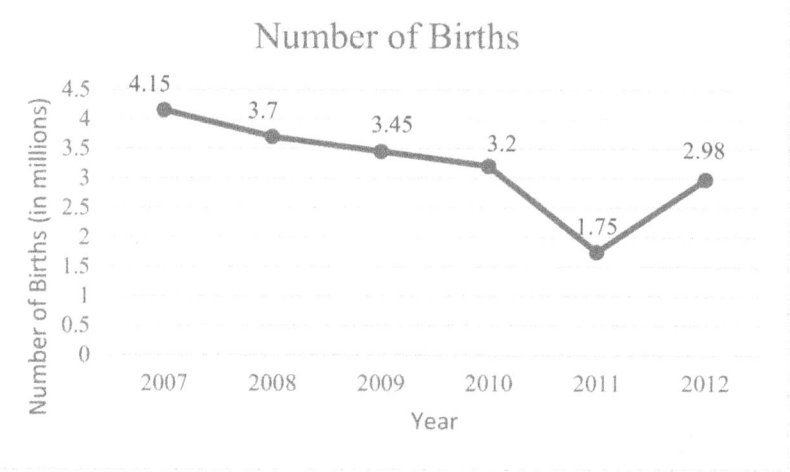

**Stem–And–Leaf Plot**

1)

| Stem | leaf |
|---|---|
| 2 | 0 1 4 6 7 9 |
| 3 | 0 6 7 |
| 5 | 1 3 5 |

2)

| Stem | leaf |
|---|---|
| 1 | 1 4 8 9 |
| 5 | 9 9 |
| 6 | 1 5 5 6 8 9 |

3)

| Stem | leaf |
|---|---|
| 5 | 4 5 9 |
| 6 | 3 6 8 |
| 11 | 2 9 |
| 12 | 1 3 5 8 |

# ACT Aspire Subject Test Mathematics Grade 8

4)

| Stem | leaf |
|------|------|
| 3    | 2 6 9 |
| 5    | 1 6 6 6 |
| 8    | 4 5 9 |
| 10   | 0 6 7 |

5)

| Stem | leaf |
|------|------|
| 1    | 1 3 6 7 9 |
| 3    | 0  3 5 5 |
| 8    | 1 6 7 9 |

6)

| Stem | leaf |
|------|------|
| 2    | 2 1 5 9 |
| 6    | 0 2 4 7 |
| 9    | 2 3 5 8 |

## Pie Graph

1) $5,600
2) $784
3) $1,288
4) $2,016
5) $\frac{1}{2}$

## Probability Problems

1) $\frac{3}{5}$
2) $\frac{1}{4}, \frac{2}{3}$
3) $\frac{1}{4}$
4) $\frac{1}{13}$
5) $\frac{7}{15}$
6) $\frac{1}{2}$
7) $\frac{1}{2}$
8) $\frac{1}{8}$
9) 0

## Factorials

1) 25
2) 122
3) 4
4) 114
5) 724
6) 9
7) 64
8) 324
9) 26
10) 90
11) 30
12) 6
13) 210
14) $n(n-1)(n-2)$
15) $(n+1)(n+2)$
16) 32
17) $5(n+2)$
18) 19.25
19) 26
20) 3
21) 16
22) $\frac{11}{13}$
23) 3,600
24) 15
25) $\frac{24}{5n(5n-1)}$
26) $\frac{n}{(n+2)}$
27) $\frac{1}{n(n-1)(n+2)}$

## Combinations and Permutations

1) 720
2) 240
3) 72
4) 126
5) 5,040
6) 24
7) 12
8) 4,920
9) 120
10) 120
11) 24
12) 120
13) 720
14) 5,040
15) 120
16) 42

WWW.MathNotion.Com

ACT Aspire Subject Test Mathematics Grade 8

# Chapter 12 : ACT Aspire Math Practice Tests

## Time to Test

**Time to refine your skill with a practice examination**

Take a REAL ACT Aspire Mathematics test to simulate the test day experience. After you've finished, score your test using the answer key.

**Before You Start**

- You'll need a pencil and scratch papers to take the test.
- There are two types are questions:
  - **Multiple choice questions:** for each of these questions, there are four possible answers. Choose which one is best.
  - **Grid-ins questions:** for these questions, write your answer in the box provided.
- It's okay to guess. You won't lose any points if you're wrong.
- After you've finished the test, review the answer key to see where you went wrong.

**Good Luck!**

# ACT Aspire Subject Test Mathematics Grade 8

## ACT Aspire GRADE 8 MAHEMATICS REFRENCE MATERIALS

**Linear Equations**

Slope-intercept form $\qquad y = mx + b$

Slope of a line $\qquad m = \dfrac{y_2 - y_1}{x_2 - x_1}$

**Circumference**

Circle $\qquad C = 2\pi r \qquad$ or $\qquad C = \pi d$

**Area**

Triangle $\qquad A = \dfrac{1}{2}bh$

Rectangle or Parallelogram $\qquad A = bh$

Trapezoid $\qquad A = \dfrac{1}{2}h(b_1 + b_2)$

Circle $\qquad A = \pi r^2$

**Surface Area**

|  | Lateral | Total |
|---|---|---|
| Prism | $S = ph$ | $S = ph + 2B$ |
| Cylinder | $S = 2\pi rh$ | $S = 2\pi rh + 2\pi r^2$ |

**Volume**

Prism or cylinder $\qquad V = Bh$

Pyramid or Cone $\qquad V = \dfrac{1}{3}Bh$

Sphere $\qquad V = \dfrac{4}{3}\pi r^3$

**Additional Information**

Pythagorean theorem $\qquad a^2 + b^2 = c^2$

Simple interest $\qquad I = prt$

Compound interest $\qquad I = p(1 + r)^t$

ACT Aspire Subject Test Mathematics Grade 8

# ACT Aspire Practice Test 1

# Mathematics

# GRADE 8

- **53 questions:** 33 multiple choice questions and 20 grid-ins questions.
- **Total time for this test:** 60 Minutes
- **You may use a calculator for this test.**

**Administered** *Month Year*

# ACT Aspire Subject Test Mathematics Grade 8

1) Right triangle ABC has two legs of lengths 15 cm (AB) and 20 cm (AC). What is the length of the third side (BC)?

   A. 25 cm

   B. 20 cm

   C. 30 cm

   D. 35 cm

2) When a number is subtracted from 52 and the difference is divided by that number, the result is 3. What is the value of the number?

   Write your answer in the box below.

   [ ]

3) A rope weighs 450 grams per meter of length. What is the weight in kilograms of 11.6 meters of this rope? (1 kilograms = 1000 grams)

   A. 5.22

   B. 0.522

   C. 52.2

   D. 5,220

4) In a party, 22 soft drinks are required for every 15 guests. If there are 225 guests, how many soft drinks is required?

   A. 240

   B. 440

   C. 330

   D. 430

5) A chemical solution contains 6% alcohol. If there is 15.84 ml of alcohol, what is the volume of the solution?

   Write your answer in the box below.

   [ ]

# ACT Aspire Subject Test Mathematics Grade 8

6) A bank is offering 3.95% simple interest on a savings account. If you deposit $8,000, how much interest will you earn in five years?

Write your answer in the box below.

7) What is the area of the shaded region?

A. 55

B. 66

C. 70

D. 85

8) The ratio of boys to girls in a school is 4:7. If there are 528 students in a school, how many boys are in the school.

Write your answer in the box below.

9) In two successive years, the population of a town is increased by 14% and 35%. What percent of the population is increased after two years?

A. 5.39%      C. 153.9%

B. 53.9%      D. 99%

# ACT Aspire Subject Test Mathematics Grade 8

10) Which graph shows a non–proportional linear relationship between *x* and *y*?

A.

B.

C.

D.

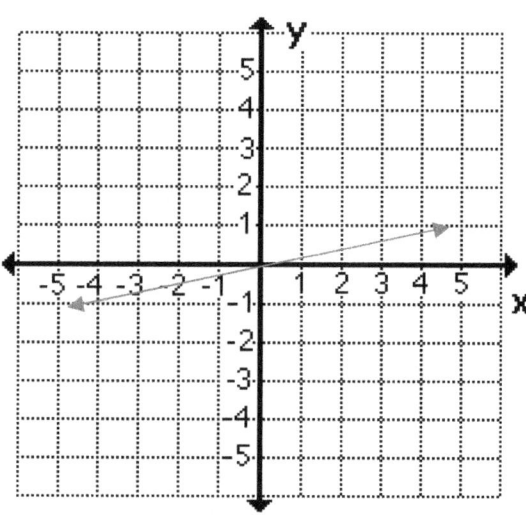

11) Five years ago, Amy was four times as old as Mike was. If Mike is 9 years old now, how old is Amy?

A. 23

B. 25

C. 21

D. 20

# ACT Aspire Subject Test Mathematics Grade 8

12) What is the value of $|-51+25|-|-3(-9)|$?

Write your answer in the box below.

☐

13) What is the solution of the following system of equations?

$$\begin{cases} \dfrac{x}{6}+\dfrac{y}{3}=2 \\ \dfrac{-2y}{3}-x=-8 \end{cases}$$

A. $x=3, y=6$

B. $x=3, y=5$

C. $x=6, y=5$

D. $x=6, y=3$

14) If a gas tank can hold 51 gallons, how many gallons does it contain when it is $\dfrac{15}{17}$ full?

Write your answer in the box below.

☐

15) What is the length of BC in the following figure if AB = 3, DF = 5 and BD = 32?

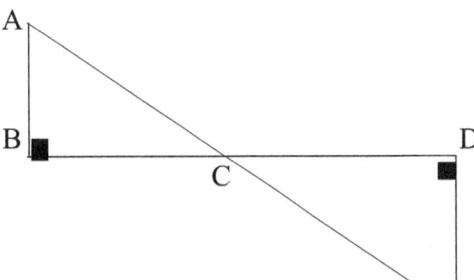

Write your answer in the box below.

☐

# ACT Aspire Subject Test Mathematics Grade 8

16) In the $xy$-plane, the point $(3, 7)$ and $(2, 3)$ are online A. Which of the following equations of lines is parallel to line A?

   A. $Y - 4X = 5$

   B. $Y = \frac{1X}{4} + \frac{7}{3}$

   C. $Y = -4X + 7$

   D. $-8y - 5x = 11$

17) In the rectangle below if $y > 9$ cm and the area of rectangle is 44 cm² and the perimeter of the rectangle is 30 cm, what is the value of $x$ and $y$ respectively?

   A. 4, 13

   B. 4, 11

   C. 6, 13

   D. 3, 11

18) A football team had $34,000 to spend on supplies. The team spent $26,000 on new balls. New sport shoes cost $220 each. Which of the following inequalities represent the number of new shoes the team can purchase?

   A. $220x + 8{,}000 \leq 34{,}000$

   B. $220x + 26{,}000 \leq 34{,}000$

   C. $8{,}000 + 220x \geq 34{,}000$

   D. $26{,}000 + 220x \geq 34{,}000$

# ACT Aspire Subject Test Mathematics Grade 8

19) A $85 shirt now selling for $51 is discounted by what percent?

　　A. 65 %　　　　　　　　　　　　C. 60 %

　　B. 35 %　　　　　　　　　　　　D. 40 %

20) How much interest is earned on a principal of $20,000 invested at an interest rate of 5% for six years?

　　　　　　　Write your answer in the box below.

　　　　　　　[ ]

21) A swimming pool holds 6,000 cubic feet of water. The swimming pool is 20 feet long and 60 feet wide. How deep is the swimming pool?

　　　　　　　Write your answer in the box below.

　　　　　　　[ ]

22) The price of a car was $40,000 in 2014, $34,000 in 2015 and $28,900 in 2016. What is the rate of depreciation of the price of car per year?

　　A. 12 %　　　　　　　　　　　　C. 15 %

　　B. 25 %　　　　　　　　　　　　D. 20%

# ACT Aspire Subject Test Mathematics Grade 8

23) The Jackson Library is ordering some bookshelves. If $x$ is the number of bookshelves the library wants to order, which each cost $90 and there is a one-time delivery charge of $850, which of the following represents the total cost, in dollar, per bookshelf?

   A. $90x + 850$

   B. $90 + 850x$

   C. $\dfrac{90x+850}{x}$

   D. $\dfrac{90x+850}{90}$

24) The following table represents the value of $x$ and function $f(x)$. Which of the following could be the equation of the function $f(x)$?

   A. $f(x) = x^2 - 11$

   B. $f(x) = 6x + 2$

   C. $f(x) = \sqrt{5x - 2}$

   D. $f(x) = 2\sqrt{x} + 5$

| $x$ | $f(x)$ |
|---|---|
| 1 | 7 |
| 4 | 9 |
| 9 | 11 |
| 16 | 13 |

25) The sum of seven different negative integers is $-77$. If the smallest of these integers is $-14$, what is the largest possible value of one of the other seven integers?

   A. $-10$            C. $-8$

   B. $-12$            D. $-7$

WWW.MathNotion.Com

# ACT Aspire Subject Test Mathematics Grade 8

26) The circle graph below shows all Mr. Wilson's expenses for last month. If he spent $960 on his car, how much did he spend for his rent?

A. $2,278

B. $2,728

C. $1,278

D. $1,728

**Mr. Wilson's Monthly Expenses**

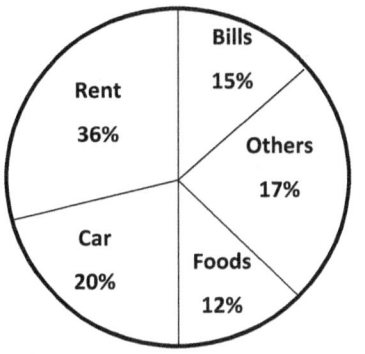

27) In the following figure, point M lies on the line A, what is the value of $y$ if $x = 20$?

Write your answer in the box below.

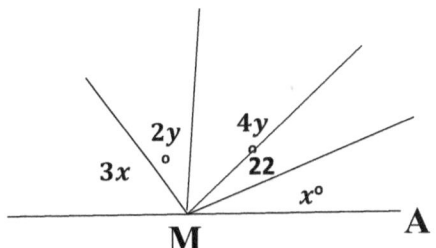

28) What is the smallest integer whose square root is greater than 11?

A. 49

B. 122

C. 100

D. 130

# ACT Aspire Subject Test Mathematics Grade 8

29) What is the area of the shaded region if the diameter of the bigger circle is 26 inches and the diameter of the smaller circle is 14 inches?

A. $120\pi$

B. $42\pi$

C. $116\pi$

D. $95\pi$

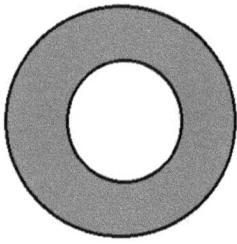

30) What is the sum of $\sqrt{2x+4}$ and $\sqrt{x}-3$ when $\sqrt{x}=-4$ ?

Write your answer in the box below.

31) What is the area of an isosceles right triangle that has one leg that measures 6 cm?

Write your answer in the box below.

32) If $x$ is directly proportional to the square of $y$, and $y = 2$ when $x = 16$, then if $x = 484$, $y = ?$

A. 7

B. 11

C. 13

D. 19

# ACT Aspire Subject Test Mathematics Grade 8

33) What is the value of $x$ in the following figure?

Write your answer in the box below.

34) Jack earns $1,000 for his first 40 hours of work in a week and is then paid 1.6 times his regular hourly rate for any additional hours. This week, Jack needs $1,600 to pay his rent, bills and other expenses. How many hours must he work to make enough money in this week?

A. 55

B. 65

C. 15

D. 25

## Questions 35, 36 and 37 are based on the following data

Types of air pollutions in 10 cities of a country

| Type of Pollution | Number of Cities | | | | | | | | | |
|---|---|---|---|---|---|---|---|---|---|---|
| A | | | | | | | | | | |
| B | | | | | | | | | | |
| C | | | | | | | | | | |
| D | | | | | | | | | | |
| E | | | | | | | | | | |
| | 1 | 2 | 3 | 4 | 5 | 6 | 7 | 8 | 9 | 10 |

# ACT Aspire Subject Test Mathematics Grade 8

35) If $a$ is the mean (average) of the number of cities in each pollution type category, $b$ is the mode, and $c$ is the median of the number of cities in each pollution type category, then which of the following must be true?

A. $b < c < a$

B. $c < b < a$

C. $a = c$

D. $a < c = b$

36) How many cities should be added to type of pollutions C until the ratio of cities in type of pollution C to cities in type of pollution D will be 1.25?

A. 8

B. 6

C. 10

D. 4

37) What percent of cities are in the type of pollution B, C, and E respectively?

A. 40%, 120%, 50%

B. 1.30%, 1.20%, 1.50%

C. 1.40%, 1.20%, 1.50%

D. 40%, 20%, 50%

38) In the following right triangle, if the sides AB and AC become triple longer, what will be the ratio of the perimeter of the triangle to its area?

A. $\frac{1}{3}$

B. 6

C. $\frac{2}{9}$

D. 9

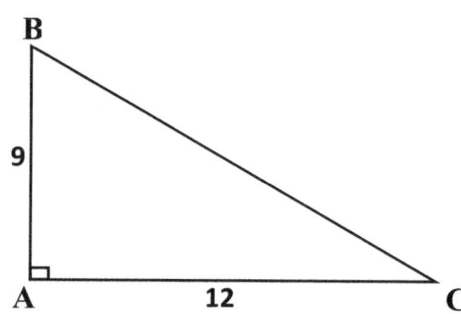

# ACT Aspire Subject Test Mathematics Grade 8

39) The capacity of a red box is 20% bigger than the capacity of a blue box. If the red box can hold 60 equal sized books, how many of the same books can the blue box hold?

A. 40

B. 50

C. 35

D. 60

40) A taxi driver earns $20 per hour work. If he works 13 hours a day, and he uses 6-liters Petrol in 3 hours with price $4.15 for 1-liter. How much money does he earn in one day?

Write your answer in the box below.

41) In the figure, *MN* is 54 cm. How long is MO?

A. 18 cm

B. 15 cm

C. 27 cm

D. 12 cm

42) Four ninth of 63 is equal to $\frac{7}{11}$ of what number?

Write your answer in the box below.

43) Which value of $x$ makes the following inequality true?

$$\frac{4}{25} \leq x < 22\%$$

A. 0.14

B. $\frac{6}{31}$

C. $\sqrt{0.044}$

D. 0.018

44) Find all values of $x$ for which $4x^2 + 10x - 6 = 0$

Write your answer in the box below.

45) $|23 - (21 \div |5 - 8|)|$ ?

Write your answer in the box below.

46) A tree 48 feet tall casts a shadow 18 feet long. Jack is 8 feet tall. How long is Jack's shadow?

A. 3 ft

B. 3.25 ft

C. 4.5 ft

D. 9 ft

# ACT Aspire Subject Test Mathematics Grade 8

47) The rectangle ABCD on the coordinate grid is translated 3 units down and 5 unit to the left.

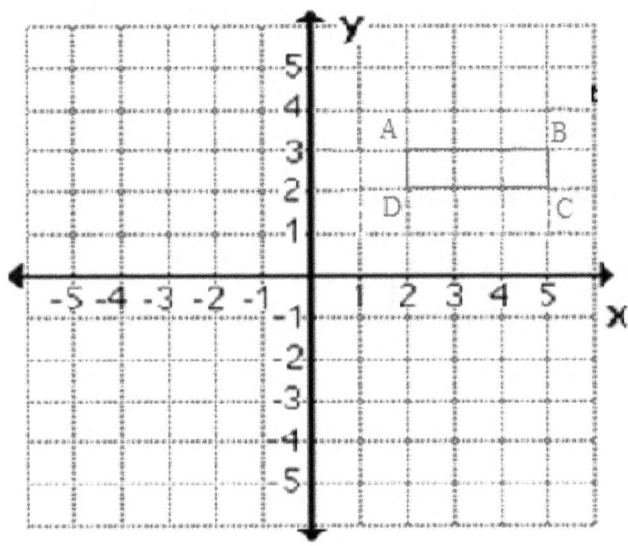

Which of the following describes this transformation?

A. $(x, y) \Rightarrow (x + 5, y - 3)$

B. $(x, y) \Rightarrow (x - 5, y - 3)$

C. $(x, y) \Rightarrow (x - 3, y + 5)$

D. $(x, y) \Rightarrow (x + 5, y + 3)$

48) If $(5.1 + 2.6 + 3.3) \times x = 2x + 18$, then what is the value of $x$?

Write your answer in the box below.

# ACT Aspire Subject Test Mathematics Grade 8

49) What is the value of $x$ in the following equation?

$$x^2 + 15 = 96$$

A. $\pm 4$     C. $\pm 8$

B. $\pm 9$     D. $\pm 11$

50) If 9 garbage trucks can collect the trash of 21 homes in a day. How many trucks are needed to collect in 140 houses?

   A. 30

   B. 45

   C. 80

   D. 60

51) The base of a triangle is 12 feet, and the interior angles are 45-45-90. What is its area?

Write your answer in the box below.

52) If $27 + x^{\frac{1}{2}} = 34$, then what is the value of $2 \times x$?

Write your answer in the box below.

# ACT Aspire Subject Test Mathematics Grade 8

53) Which set of ordered pairs represents $y$ as a function of $x$?

A. $\{(3, -2), (5, 5), (6, -8), (3, 7)\}$

B. $\{(9, 11), (3, -12), (4, 7), (2, 6)\}$

C. $\{(3, 11), (6, 8), (6, 12), (8, 14)\}$

D. $\{(6, 3), (5, 7), (5, 8), (2, 1)\}$

**STOP**

**This is the End of Test 1.**

"You may check your work on this Test if you still have time".

ACT Aspire Subject Test Mathematics Grade 8

# ACT Aspire Practice Test 2

# Mathematics

# GRADE 8

- **53 questions:** 33 multiple choice questions and 20 grid-ins questions.
- **Total time for this test:** 60 Minutes.
- You may use a calculator for this test.

**Administered** Month Year

**ACT Aspire Subject Test Mathematics Grade 8**

1) The area of a circle is 81 π. Which of the following can be the circumference of the circle?

   A. 36 π

   B. 9 π

   C. 18 π

   D. 16 π

2) You can buy 12 cans of green beans at a supermarket for $3.90. How much does it cost to buy 64 cans of green beans?

   Write your answer in the box below.

3) Which of the following is the solution of the following inequality?

   $4.5x + 9 > 8x - 3 - 1.5x$

   A. $x < 6$

   B. $x > 6$

   C. $x < -6$

   D. $x > -6$

4) What is the perimeter of a square that has an area of 282.24 feet?

   Write your answer in the box below.

5) A tree 21 feet tall casts a shadow 36 feet long. Jack is 7 feet tall. How long is Jack's shadow?

   A. 18 ft

   B. 12 ft

   C. 24 ft

   D. 42 ft

# ACT Aspire Subject Test Mathematics Grade 8

6) The price of a laptop is decreased by 30% to $620. What is its original price?

   A. 58.71

   B. 885.41

   C. 85.87

   D. 885.71

7) A container holds 5.4 gallons of water when it is $\frac{9}{47}$ full. How many gallons of water does the container hold when it's full?

   A. 26

   B. 28.2

   C. 26.12

   D. 18.52

8) If $(2^a)^b = 512$, then what is the value of $a \times b$?

   Write your answer in the box below.

   ☐

9) A bag contains 15 balls: six green, four black, one blue, three red and one white. If 14 balls are removed from the bag at random, what is the probability that a white ball has been removed?

   A. $\frac{1}{15}$

   B. $\frac{1}{29}$

   C. $\frac{1}{14}$

   D. $\frac{14}{15}$

10) What is the $x$-intercept of the line with equation $5x - 8y = 35$?

    Write your answer in the box below.

    ☐

ACT Aspire Subject Test Mathematics Grade 8

11) Which of the following expressions is equivalent to $7x^3y(6x + 3y)$?

   A. $42yx^3 - 21xy$

   B. $42yx^4 + 21x^3y^2$

   C. $-42yx^4 + 21x^3y^2$

   D. $-42yx^4 - 21x^3y^2$

12) A soccer team played 160 games and won 42 percent of them. How many games did the team win?

   Write your answer in the box below.

13) The equation of a line is given as: $y = -3x + 2$. Which of the following points does not lie on the line?

   A. $(1, -1)$

   B. $(3, -7)$

   C. $(-2, 8)$

   D. $(-3, 7)$

14) The perimeter of the trapezoid below is 64 cm. What is its area?

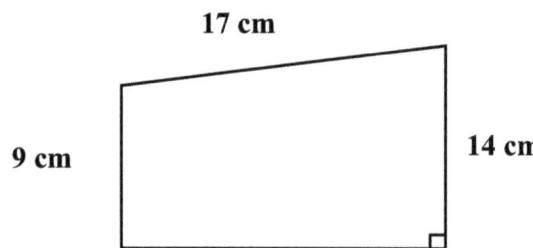

Write your answer in the box below.

15) Which graph does not represent $y$ as a function of $x$?

A.

B.

C.

D.
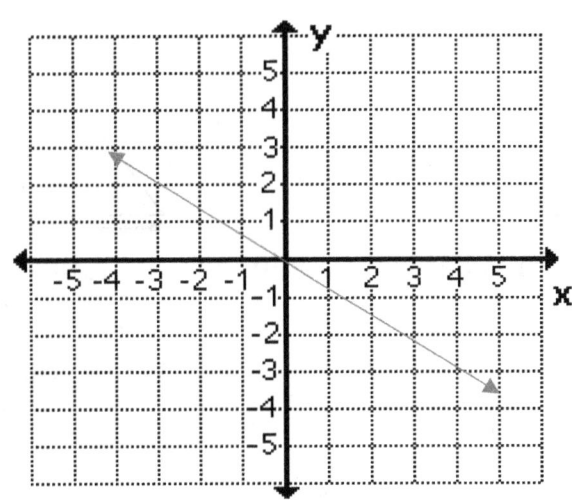

16) Which of the following is equivalent to $-25 < -7x - 4 < -4$ ?

A. $-3 < x < 0$

B. $0 < x < 3$

C. $-3 < x < 0$

D. $-3 < x < 3$

# ACT Aspire Subject Test Mathematics Grade 8

17) A bank is offering 1.95% simple interest on a savings account. If you deposit $24,000, how much interest will you earn in five years?

A. $1.430

C. $1,520

B. $2,340

D. $234

18) Joe scored 33 out of 43 marks in Algebra, 26 out of 51 marks in science and 18 out of 29 marks in mathematics. In which subject his percentage of marks is best?

A. Algebra

C. Mathematics and Science

B. Mathematics

D. Science

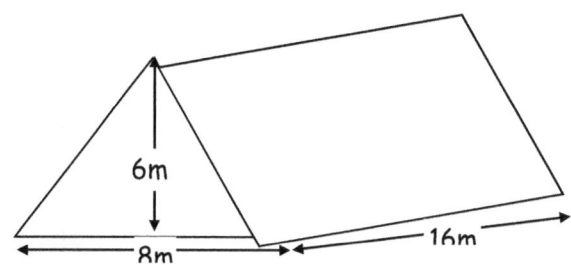

19) What is the volume of the following triangular prism?

Write your answer in the box below.

20) The marked price of a computer is E Euro. Its price decreased by 26% in March and later increased by 7 % in April. What is the final price of the computer in E Euro?

A. 0.702 E

C. 0.079 E

B. 0.072E

D. 0.792 E

# ACT Aspire Subject Test Mathematics Grade 8

21) Find the slope–intercept form of the graph $3x - 5y = -14$

A. $y = 3x + \frac{5}{14}$

B. $y = \frac{1}{5}x + \frac{3}{5}$

C. $y = \frac{3}{5}x + 2\frac{4}{5}$

D. $y = \frac{3}{5}x - 2\frac{3}{4}$

22) Triangle ABC is graphed on a coordinate grid with vertices at A (8, 3), B(– 2, 6) and C (4, 7). Triangle ABC is reflected over x axes to create triangle A'B'C'. Which order pair represents the coordinate of C'?

A. (7,4)

B. (−4, −7)

C. (4, −7)

D. (−4, 7)

23) Which of the following point is the solution of the system of equations?

$$\begin{cases} 7x - 10y = 9 \\ 2x - 3y = 3 \end{cases}$$

A. (−3, −3)

B. (−3, 3)

C. (3, 3)

D. (3, −3)

24) A shirt costing $410 is discounted 23%. After a month, the shirt is discounted another 9%. Which of the following expressions can be used to find the selling price of the shirt?

A. (410)(0.23)(0.09)

B. (410)(0.77)(0.91)

C. (410) (0.23) − (220) (0.23)

D. (410)(0.23) − (410)(0.09)

# ACT Aspire Subject Test Mathematics Grade 8

25) What is the distance between the points $(7, -8)$ and $(2, 4)$?

Write your answer in the box below.

## Questions 26 and 27 are based on the following data

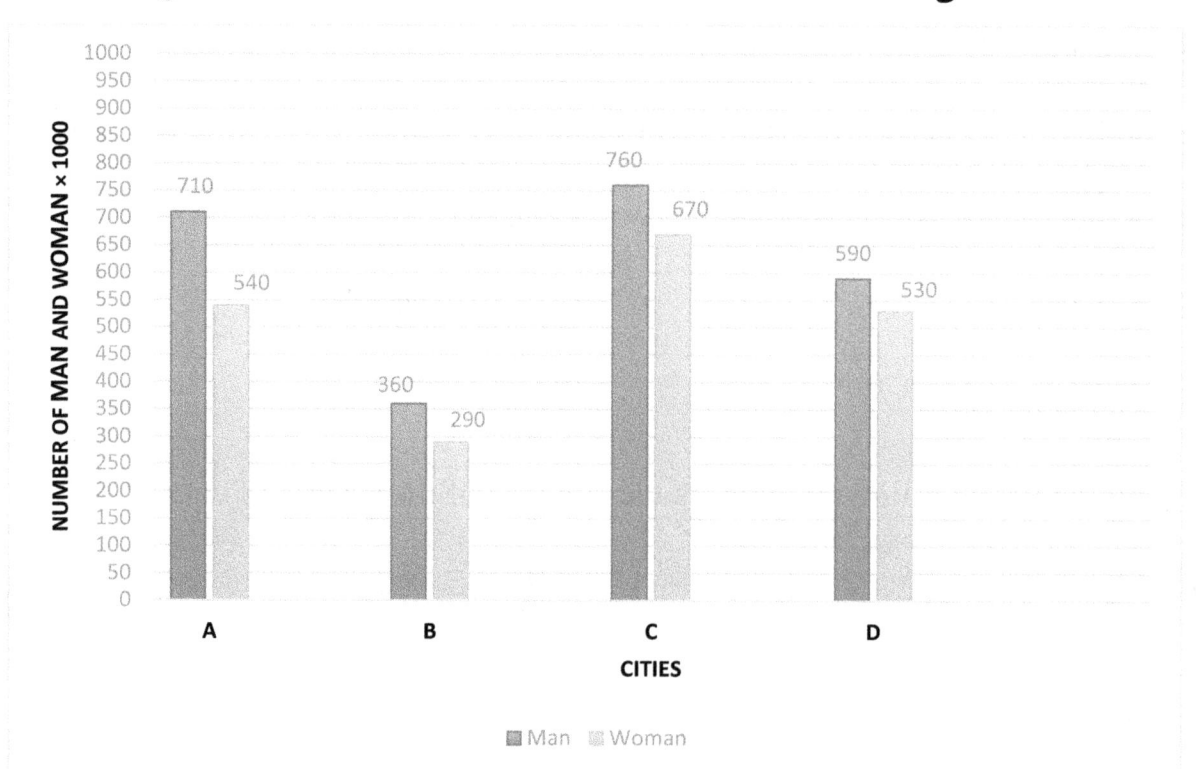

26) What's the ratio of percentage of men in city C to percentage of women in city B?

A. 0.41

B. 0.041

C. 1.19

D. 1.09

27) What's the minimum ratio of woman to man in the four cities?

A. 0.90

B. 0.67

C. 0.88

D. 0.76

# ACT Aspire Subject Test Mathematics Grade 8

28) Line m passes through the point $(3, -7)$. Which of the following **CANNOT** be the equation of line m?

A. $y = -7$

B. $y = -4x + 5$

C. $y = 8 - 3x$

D. $y = -5x + 8$

29) The sum three numbers are 46. If another number is added to these four numbers, the average of the four numbers is 16. What is the fourth number?

Write your answer in the box below.

☐

30) David owed $5,048. After making 19 payments of $167 each, how much did he have left to pay?

A. $2,578

B. $1,875

C. $1,755

D. $1,857

31) Which of the following lines is parallel to the graph of $y = 9x - 4$ ?

A. $9x - 4y = 4$

B. $9x - 7y = 0$

C. $9x - y = -4$

D. $4x + y = 4$

32) The average of six consecutive numbers is 39. What is the smallest number?

A. 36

B. 32

C. 32.5

D. 36.5

33) The price of a laptop is decreased by 40% to $540. What is its original price?

Write your answer in the box below.

☐

34) Three seventh of 42 is equal to $\frac{2}{3}$ of what number?

Write your answer in the box below.

☐

35) The average weight of 13 girls in a class is 35 kg and the average weight of 22 boys in the same class is 45 kg. What is the average weight of all the 35 students in that class?

A. 38.80 Kg

B. 39 Kg

C. 46 Kg

D. 41.28 Kg

36) Liam's average (arithmetic mean) on three mathematics tests is 56. What should Liam's score be on the next test to have an overall of 64 for all the tests?

Write your answer in the box below.

☐

37) Which of the following equations has a graph that is a straight line?

A. $y = 5x^2 - 2$

B. $y^2 - 3x^2 = 2$

C. $x - y = 3$

D. $4x^2 + y = 3$

# ACT Aspire Subject Test Mathematics Grade 8

38) An angle is equal to one fourth of its supplement. What is the measure of that angle?

Write your answer in the box below.

39) The radius of circle A is four times the radius of circle B. If the circumference of circle A is $32\pi$, what is the area of circle B?

Write your answer in the box below.

40) A card is drawn at random from a standard 52–card deck, what is the probability that the card is of Hearts or Spades or Hearts? (The deck includes 13 of each suit clubs, diamonds, hearts, and spades)

A. $\frac{3}{52}$      C. $\frac{1}{3}$

B. $\frac{3}{4}$       D. $\frac{1}{4}$

41) The score of Zoe was one sixth of Emma and the score of Harper was twice that of Emma. If the score of Harper was 96, what is the score of Zoe?

A. 8       C. 6

B. 12      D. 10

# ACT Aspire Subject Test Mathematics Grade 8

42) List A consists of the numbers {4, 7, 9, 13, 19}, and list B consists of the numbers {6, 15, 17, 23, 26}.

If the two lists are combined, what is the median of the combined list?

Write your answer in the box below.

43) If $f(x) = 8x^3 + 7$, what is the smallest possible value of $f(x)$?

A. 3

B. 2

C. 7

D. 8

44) If $4x + 5y = 8$ and $4x - 5y = 6$ then what is the value of $(16x^2 - 25y^2)$?

Write your answer in the box below.

45) A circular logo is enlarged to fit the lid of a jar. The new diameter is 30% larger than the original. By what percentage has the area of the logo increased?

A. 25%

B. 99%

C. 69%

D. 54%

46) $\frac{1}{8b^2} + \frac{1}{8b} = \frac{1}{4b^2}$, then $b = ?$

Write your answer in the box below.

# ACT Aspire Subject Test Mathematics Grade 8

47) $(3x + 5)(x + 7) =$

    A. $6x + 14$                                    C. $3x^2 + 13x + 35$

    B. $6x + 12x + 14$                        D. $3x^2 + 26x + 35$

48) What is the area of an isosceles right triangle with hypotenuse that measures 16 cm?

Write your answer in the box below.

49) Which of the following is a common factor of following equations?

$$x^2 - x - 24 \text{ and } x^2 - 5x - 56$$

    A. $(x - 8)$                                    C. $(x - 3)$

    B. $(x + 8)$                                    D. $(x + 3)$

50) If $x \blacksquare y = \sqrt{x^2 + 8y}$, what is the value of $3 \blacksquare 5$?

Write your answer in the box below.

51) What is the number of cubic feet of soil needed for a flower box 6 feet long, 9 inches wide, and 4 feet deep?

    A. 18 cubic feet                             C. $\frac{3}{4}$ cubic feet

    B. 16 cubic feet                             D. 24 cubic feet

# ACT Aspire Subject Test Mathematics Grade 8

52) $6x^5y^2 + 3x^4y^7 - (3x^5y^2 + 9x^4y^7) = $ ___

   A. $-3x^5y^2 + 9x^4y^7$

   B. $6x^5y^4 + 9x^4y^7$

   C. $-6x^4y^7$

   D. $3x^5y^2 - 9x^4y^7$

53) Calculate the area of a parallelogram with a base of 5 feet and height of 6.9 feet.

   Write your answer in the box below.

   [ ]

**STOP**

This is the End of Test 2.

"You may check your work on this Test if you still have time".

# Chapter 13 : Answers and Explanations

# ACT Aspire Practice Tests
## Answer Key

✻ Now, it's time to review your results to see where you went wrong and what areas you need to improve!

## ACT Aspire - Mathematics

### Practice Test - 1

| | | | | | |
|---|---|---|---|---|---|
| 1 | A | 19 | D | 37 | D |
| 2 | 13 | 20 | 6,000 | 38 | C |
| 3 | A | 21 | 5 | 39 | B |
| 4 | C | 22 | C | 40 | 160.4 |
| 5 | 264 | 23 | C | 41 | A |
| 6 | 1,580 | 24 | D | 42 | 44 |
| 7 | A | 25 | C | 43 | B |
| 8 | 192 | 26 | D | 44 | $-3, \frac{1}{2}$ |
| 9 | B | 27 | 13 | 45 | 16 |
| 10 | B | 28 | B | 46 | A |
| 11 | C | 29 | A | 47 | B |
| 12 | −1 | 30 | 7 | 48 | 2 |
| 13 | D | 31 | 18 | 49 | B |
| 14 | 45 | 32 | B | 50 | D |
| 15 | 12 | 33 | 61° | 51 | 72 |
| 16 | A | 34 | A | 52 | 98 |
| 17 | B | 35 | C | 53 | B |
| 18 | B | 36 | A | | |

### Practice Test - 2

| | | | | | |
|---|---|---|---|---|---|
| 1 | C | 19 | $384\,m^3$ | 37 | C |
| 2 | 20.8 | 20 | D | 38 | 36 |
| 3 | A | 21 | C | 39 | $16\pi$ |
| 4 | 67.2 | 22 | C | 40 | B |
| 5 | B | 23 | A | 41 | A |
| 6 | D | 24 | B | 42 | 14 |
| 7 | B | 25 | 13 | 43 | C |
| 8 | 9 | 26 | C | 44 | 48 |
| 9 | A | 27 | D | 45 | C |
| 10 | 7 | 28 | C | 46 | 1 |
| 11 | B | 29 | 18 | 47 | D |
| 12 | 67.2 | 30 | B | 48 | 64 |
| 13 | D | 31 | C | 49 | A |
| 14 | 276 | 32 | D | 50 | 7 |
| 15 | A | 33 | 900 | 51 | A |
| 16 | B | 34 | 27 | 52 | D |
| 17 | B | 35 | D | 53 | 34.5 |
| 18 | A | 36 | 88 | | |

ACT Aspire Subject Test Mathematics Grade 8

# Practice Test 1
# ACT Aspire - Mathematics
# Answers and Explanations

**1) Answer: A.**

Use Pythagorean Theorem: $a^2 + b^2 = c^2$

$15^2 + 20^2 = C^2 \Rightarrow 225 + 400 = C^2 \Rightarrow 625 = c^2 \Rightarrow c = 25$

**2) Answer: 13.**

Let $x$ be the number. Write the equation and solve for $x$.

$\frac{(52-x)}{x} = 3$ (cross multiply)

$(52 - x) = 3x$, then add $x$ both sides. $52 = 4x$, now divide both sides by 4. $\Rightarrow x = 13$.

**3) Answer: A.**

The weight of 11.6 meters of this rope is: $11.6 \times 450g = 5,220g$

1 kg = 1,000 g, therefore, $5,220\ g \div 1000 = 5.22 kg$

**4) Answer: C.**

Let $x$ be the number of soft drinks for 225 guests. Write the proportion and solve for $x$.

$\frac{22\ \text{soft drinks}}{15\ \text{guests}} = \frac{x}{225\ \text{guests}} \Rightarrow x = \frac{225 \times 22}{15} \Rightarrow x = 330$

**5) Answer: 264.**

6% of the volume of the solution is alcohol. Let $x$ be the volume of the solution. Then:

6% of $x = 15.84$ ml

$0.06x = 15.84 \Rightarrow \frac{6x}{100} = \frac{158.4}{10}$ cross multiply; $60x = 15,840 \Rightarrow$ (devide by 60) $x = 264$

**6) Answer: 1,580.**

Use simple interest formula: $I = prt$ ; (I=interest, p=principal, r=rate, t=time)

$I = (8,000)(0.0395)(5) = 1,580$.

**7) Answer: A.**

Use the area of rectangle formula ($s = a \times b$).

To find area of the shaded region subtract smaller rectangle from bigger rectangle.

$S_1 - S_2 = (7\ ft \times 10 ft) - (3 ft \times 5 ft) \Rightarrow S_1 - S_2 = 55 ft$.

# ACT Aspire Subject Test Mathematics Grade 8

**8) Answer: 192.**

The ratio of boy to girls is 4:7. Therefore, there are 4 boys out of 11 students. To find the answer, first divide the total number of students by 11, then multiply the result by 4.

$528 \div 11 = 48 \Rightarrow 48 \times 4 = 192$

**9) Answer: B.**

the population is increased by 14% and 35%. 14% increase changes the population to 114% of original population.

For the second increase, multiply the result by 135%.

$(1.14) \times (1.35) = 1.539 = 153.9\%$

53.9 percent of the population is increased after two years.

**10) Answer: B.**

A linear equation is a relationship between two variables, $x$ and $y$, that can be put in the form $y = mx + b$.

A non-proportional linear relationship takes on the form $y = mx + b$, where $b \neq 0$ and its graph is a line that does not cross through the origin.

**11) Answer: C.**

Five years ago, Amy was four times as old as Mike. Mike is 9 years now. Therefore, 5 years ago Mike was 4 years.

Five years ago, Amy was: $A = 4 \times 4 = 16$

Now Amy is 21 years old: $16 + 5 = 21$

**12) Answer: −1.**

$|-51 + 25| - |-3(-9)| = |-26| - |-27| = 26 - 27 = -1$

**13) Answer: D.**

$\begin{cases} \frac{x}{6} + \frac{y}{3} = 2 \\ \frac{-2y}{3} - x = -8 \end{cases}$ → Multiply the top equation by 6. Then,

$\begin{cases} x + 2y = 12 \\ \frac{-2y}{3} - x = -8 \end{cases}$ → Add two equations.

$\frac{4}{3}y = 4 \to y = 3$, plug in the value of y into the first equation→ $x = 6$

# ACT Aspire Subject Test Mathematics Grade 8

**14) Answer: 45.**

$\frac{15}{17} \times 51 = \frac{765}{17} = 45$

**15) Answer: 12.**

Two triangles $\triangle ABC$ and $\triangle CD$ are similar. Then:

$\frac{AB}{DF} = \frac{BC}{CD} \to \frac{3}{5} = \frac{x}{32-x} \to 96 - 3x = 5x \to 8x = 96 \to x = 12$

**16) Answer: A.**

The slop of line A is: $m = \frac{y_2 - y_1}{x_2 - x_1} = \frac{3-7}{2-3} = 4$

Parallel lines have the same slope and only choice A has slope of 4.

$(y - 4x = 5 \Rightarrow y = 4x + 5)$.

**17) Answer: B.**

The perimeter of the rectangle is: $2x + 2y = 30 \to x + y = 15 \to x = 15 - y$

The area of the rectangle is: $x \times y = 44 \to (15 - y)(y) = 44 \to y^2 - 15y + 44 = 0$

Solve the quadratic equation by factoring method.

$(y - 4)(y - 11) = 0 \to y = 4$ (Unacceptable, because $y$ must be greater than 9) or $y = 11$; If $y = 11 \to x = 15 - y \to x = 15 - 11 \to x = 4$

**18) Answer: B.**

Let $x$ be the number of new shoes the team can purchase. Therefore, the team can purchase $220x$.

The team had $34,000 and spent $26,000. Now the team can spend on new shoes $8,000 at most. Now, write the inequality: $220x + 26,000 \leq 34,000$

**19) Answer: D.**

Use the formula for Percent of Change:

$\frac{\text{New Value} - \text{Old Value}}{\text{Old Value}} \times 100\% \Rightarrow \frac{51-85}{85} \times 100\% = -40\%$

(negative sign here means that the new price is less than old price).

**20) Answer: 6,000.**

Use simple interest formula: $I = prt$ (I = interest, p = principal, r = rate, t = time)

$I = (20,000)(0.05)(6) = 6,000$

# ACT Aspire Subject Test Mathematics Grade 8

**21) Answer: 5.**

Use formula of rectangle prism volume.

V = (length) (width) (height) ⇒ 6,000 = (20) (60) (height)

⇒ height = 6,000 ÷ 1,200 = 5

**22) Answer: C.**

Use this formula: Percent of Change = $\frac{\text{New Value} - \text{Old Value}}{\text{Old Value}} \times 100\%$

$\frac{34,000 - 40,000}{40,000} \times 100\% = -15\%$ and $\frac{28,900 - 34,000}{34,000} \times 100\% = -15\%$

**23) Answer: C.**

The amount of money for $x$ bookshelf is: $90x$

Then, the total cost of all bookshelves is equal to: $90x + 850$

The total cost, in dollar, per bookshelf is:

$\frac{\text{Total cost}}{\text{number of items}} = \frac{90x + 850}{x}$

**24) Answer: D.**

A. $f(x) = x^2 - 11$   if   $x = 9 \to f(9) = 9^2 - 11 = 70 \neq 11$

B. $f(x) = 6x + 2$   if   $x = 9 \to f(9) = 6(9) + 2 = 56 \neq 11$

C. $f(x) = \sqrt{5x - 2}$   if   $x = 9 \to f(9) = \sqrt{5(9) - 2} = \sqrt{43} \neq 11$

D. $f(x) = 2\sqrt{x} + 5$   if   $x = 9 \to f(9) = 2\sqrt{9} + 5 = 11$

**25) Answer: C.**

The smallest number is $-14$.

To find the largest possible value of one of the other seven integers, we need to choose the smallest possible integers for six of them. Let $x$ be the largest number. Then:

$-77 = (-14) + (-13) + (-12) + (-11) + (-10) + (-9) + x \to -77 = -69 + x$

$\to x = -77 + 69 = -8$

**26) Answer: D.**

Let $x$ be all expenses, then $\frac{20}{100}x = \$960 \to x = \frac{100 \times \$960}{20} = \$4,800$

He spent for his rent: $\frac{36}{100} \times \$4,800 = \$1,728$

WWW.MathNotion.Com

# ACT Aspire Subject Test Mathematics Grade 8

**27) Answer: 13.**

The angles on a straight line add up to 180 degrees. Then:

$x + 22 + 4y + 2y + 3x = 180$

Then, $4x + 6y = 180 - 22 \rightarrow 4(20) + 6y = 158$

$\rightarrow 6y = 158 - 80 = 78 \rightarrow y = 13$

**28) Answer: B.**

Square root of 49 is $\sqrt{49} = 7 < 11$

Square root of 122 is $\sqrt{122} = \sqrt{121+1} > \sqrt{121} = 11$

Square root of 100 is $\sqrt{100} = 10 < 11$

Square root of 130 is $\sqrt{130} = \sqrt{121+9} > 11$

Since $\sqrt{122} < \sqrt{130}$, then the answer is B.

**29) Answer: A.**

To find the area of the shaded region subtract smaller circle from bigger circle.

S bigger − S smaller = $\pi \, (r \, bigger\,)^2 - \pi \, (r \, smaller\,)^2$

$\Rightarrow$ S bigger − S smaller = $\pi \, (13)^2 - \pi \, (7)^2 \Rightarrow 169\pi - 49\pi = 120\pi$

**30) Answer: 7.**

$\sqrt{x} = -4 \rightarrow x = 16$

then; $\sqrt{x} - 3 = \sqrt{16} - 3 = 4 - 3 = 1$ and $\sqrt{2x+4} = \sqrt{32+4} = \sqrt{36} = 6$

Then: $(\sqrt{2x+4}) + (\sqrt{x} - 3) = 6 + 1 = 7$

**31) Answer: 18.**

$a = 6 \Rightarrow$ area of the triangle is:

$= \frac{1}{2}(6 \times 6) = \frac{36}{2} = 18 \, cm^2$

Isosceles right triangle

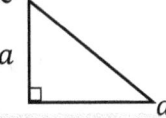

**32) Answer: B.**

$x$ is directly proportional to the square of $y$. Then: $x = cy^2$

$16 = c(2)^2 \rightarrow 16 = 4c \rightarrow c = \frac{16}{4} = 4$

The relationship between $x$ and $y$ is: $x = x = 4y^2$, $x = 484$

$484 = 4y^2 \rightarrow y^2 = \frac{484}{4} = 121 \rightarrow y = 11$

# ACT Aspire Subject Test Mathematics Grade 8

**33) Answer: 61°.**

$\alpha = 180° - 129° = 51°$

$\beta = 180° - 112° = 68°$

$x + \alpha + \beta = 180° \rightarrow x = 180° - 51° - 68° = 61°$

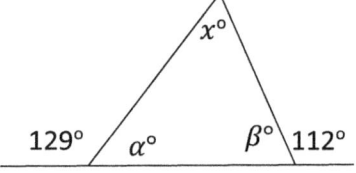

**34) Answer: A.**

The amount of money that jack earns for one hour: $\frac{\$1,000}{40} = \$25$

Number of additional hours that he works to make enough money is: $\frac{\$1,600 - \$1,000}{1.6 \times \$25} = 15$

Number of total hours is: $40 + 15 = 55$

**35) Answer: C.**

Let's find the mean (average), mode and median of the number of cities for each type of pollution.

Number of cities for each type of pollution: 6, 4, 2, 8, 5

average (mean) $= \frac{sum of terms}{number of terms} = \frac{6+4+2+8+5}{5} = \frac{25}{5} = 5$

Median is the number in the middle. To find median, first list numbers in order from smallest to largest: 2, 4, 5, 6, 8

Median of the data is 5.

Mode is the number which appears most often in a set of numbers. Therefore, there is no mode in the set of numbers.

Median = Mean, then, a=c

**36) Answer: A.**

Let the number of cities should be added to type of pollutions C be $x$.

Then: $\frac{x+2}{8} = 1.25 \rightarrow x + 2 = 8 \times 0.75 \rightarrow x + 2 = 10 \rightarrow x = 8$

**37) Answer: D.**

Percent of cities in the type of pollution B: $\frac{4}{10} \times 100 = 40\%$

Percent of cities in the type of pollution C: $\frac{2}{10} \times 100 = 20\%$

Percent of cities in the type of pollution E: $\frac{5}{10} \times 100 = 50\%$

# ACT Aspire Subject Test Mathematics Grade 8

**38) Answer: C.**

$AB = 9$, And $AC = 12$

$BC = \sqrt{9^2 + 12^2} = \sqrt{81 + 144} = \sqrt{255} = 15$

Perimeter $= 9 + 12 + 15 = 36$ ; Area $= \frac{9 \times 12}{2} = 54$

In this case, the ratio of the perimeter of the triangle to its area is: $\frac{36}{54} = \frac{2}{3}$

If the sides AB and AC become twice longer, then: $AB = 27$, And $AC = 36$

$BC = \sqrt{27^2 + 36^2} = \sqrt{729 + 1{,}296} = \sqrt{2{,}025} = 45$

Perimeter $= 27 + 36 + 45 = 108$ ; Area $= \frac{27 \times 36}{2} = 27 \times 18 = 486$

In this case the ratio of the perimeter of the triangle to its area is: $\frac{108}{486} = \frac{2}{9}$

**39) Answer: B.**

The capacity of a red box is 20% bigger than the capacity of a blue box and it can hold 60 books. Therefore, we want to find a number that 20% bigger than that number is 20. Let $x$ be that number. Then: $1.20 \times x = 60$, Divide both sides of the equation by 1.2.

Then: $x = \frac{60}{1.20} = 50$

**40) Answer: $160.4.**

$\$20 \times 13 = \$260$

Petrol use: $6 \div 3 = 2, 12 \times 2 = 24$ liters

Petrol cost: $24 \times \$4.15 = \$99.6$

Money earned: $\$260 - \$99.6 = \$160.4$

**41) Answer: A.**

The length of MN is equal to: $3x + 6x = 9x$

Then: $9x = 54 \rightarrow x = \frac{54}{9} = 6$

The length of MO is equal to: $3x = 3 \times 6 = 18$cm.

**42) Answer: 44.**

Let x be the number. Write the equation and solve for x.

$\frac{4}{9} \times 63 = \frac{7}{11}x \rightarrow \frac{4 \times 63}{9} = \frac{7x}{11}$, use cross multiplication to solve for x.

WWW.MathNotion.Com

# ACT Aspire Subject Test Mathematics Grade 8

$44 \times 63 = 7x \times 9 \Rightarrow 2,772 = 63x \Rightarrow x = 44$

**43) Answer: B.**

$\frac{4}{25} = 0.16$ and $22\% = 0.22$ therefore $x$ should be between 0.16 and 0.22

Only choice B ($\frac{6}{31} = 0.193$) is between 0.16 and 0.22

**44) Answer:** $-3, \frac{1}{2}$.

First, divide all values on both sides of the equation by 2.

Then: $2x^2 + 5x - 3 = 0$

Use quadratic formula: $ax^2 + bx + c = 0$

$x_{1,2} = \frac{-b \pm \sqrt{b^2 - 4ac}}{2a}$

$2x^2 + 5x - 3 = 0 \Rightarrow a = 2, b = 5$ and $c = -3$

$x_1 = \frac{-5 + \sqrt{5^2 - 4 \times 2 \times (-3)}}{2 \times 2} = \frac{-5 + \sqrt{25 + 24}}{4} = \frac{-5 + \sqrt{49}}{4} = \frac{1}{2}$

$x_2 = \frac{-5 - \sqrt{25 + 24}}{2 \times 2} = \frac{-5 - \sqrt{49}}{4} = -3$

**45) Answer: 16.**

Simplify: $|23 - (21 \div |5 - 8|)| = |23 - (21 \div |-3|)| = |23 - (21 \div 3)| = |23 - 7| = |16| = 16$

**46) Answer: A.**

Write a proportion and solve for the missing number.

$\frac{48}{18} = \frac{8}{x} \rightarrow 48x = 8 \times 18 = 144 \Rightarrow 48x = 144 \rightarrow x = \frac{144}{48} = 3$

**47) Answer: B.**

Translated 3 units down and 5 units to the left means:

$(x, y) \Rightarrow (x - 5, y - 3)$

**48) Answer: 2.**

$(5.1 + 2.6 + 3.3) \times x = 2x + 18 \Rightarrow 11x = 2x + 18 \Rightarrow 9x = 18$ Then $x = 2$

**49) Answer: B.**

$x^2 + 15 = 96 \Rightarrow x^2 = 96 - 15 \Rightarrow x^2 = 81 \Rightarrow x = \pm 9$

# ACT Aspire Subject Test Mathematics Grade 8

**50) Answer: D.**

Write a proportion and solve. $\frac{9}{21} = \frac{x}{140} \rightarrow x = \frac{9 \times 140}{21} = 60$

**51) Answer: 72.**

Formula of triangle area = ½ (base × height)

Since the angles are 45-45-90, then this is an isosceles triangle, meaning that the base and height of the triangle are equal.

Triangle area = ½ (base × height) = ½ (12 × 12) = 72

**52) Answer: 98.**

$x^{\frac{1}{2}}$ equals to the root of $x$. Then:

$27 + x^{\frac{1}{2}} = 34 \rightarrow 27 + \sqrt{x} = 34 \rightarrow \sqrt{x} = 7 \rightarrow x = 49$

$x = 49$ and $2 \times x$ equals: $2 \times 49 = 98$

**53) Answer: B.**

A set of ordered pairs represents $y$ as a function of $x$ if: $x_1 = x_2 \rightarrow y_1 = y_2$

In choice A: $(3, -2)$ and $(3, 7)$ are ordered pairs with same x and different y, therefore y isn't a function of $x$.

In choice C: $(6, 8)$ and $(6, 12)$ are ordered pairs with same x and different y, therefore y isn't a function of $x$.

In choice D: $(5, 7)$ and $(5, 8)$ are ordered pairs with same x and different y, therefore y isn't a function of $x$.

# ACT Aspire Subject Test Mathematics Grade 8

# Practice Test 2
# ACT Aspire - Mathematics
# Answers and Explanations

**1) Answer: C.**

Use the formula for area of circles.

Area = $\pi r^2 \Rightarrow 81\pi = \pi r^2 \Rightarrow 81 = r^2 \Rightarrow r = 9$

Radius of the circle is 9. Now, use the circumference formula:

Circumference = $2\pi r = 2\pi (9) = 18 \pi$

**2) Answer: $20.8**

Let $x$ be the number of cans. Write the proportion and solve for $x$.

$\frac{12 \text{ cans}}{\$ 3.90} = \frac{64 \text{ cans}}{x} \Rightarrow x = \frac{3.90 \times 64}{12} \Rightarrow x = \$20.8$

**3) Answer: A.**

$4.5x + 9 > 8x - 3 - 1.5x \rightarrow$ Combine like terms:

$4.5x + 9 > 6.5x - 3 \rightarrow$ Subtract $4.5x$ from both sides: $9 > 2x - 3$

Add 3 both sides of the inequality.

$12 > 2x$, Divide both sides by 2. $\Rightarrow \frac{12}{2} > x \rightarrow x < 6$

**4) Answer: 67.2 feet.**

Area of a square: $S = a^2 \Rightarrow 282.24 = a^2 \Rightarrow a = 16.8$

Perimeter of a square: $P = 4a \Rightarrow P = 4 \times 16.8 \Rightarrow P = 67.2$

**5) Answer: B.**

Write the proportion and solve for the missing number.

$\frac{21}{36} = \frac{7}{x} \rightarrow 21x = 7 \times 36 = 252 \rightarrow 21x = 252 \rightarrow x = \frac{252}{21} = 12$

**6) Answer: D.**

Let $x$ be the original price.

If the price of a laptop is decreased by 30% to $620, then:

$70 \% \; of \; x = 620 \Rightarrow 0.70x = 620 \Rightarrow x = 620 \div 0.70 = 885.71$

## ACT Aspire Subject Test Mathematics Grade 8

**7) Answer: B.**

let $x$ be number of gallons of water the container holds when it is full.

Then; $\frac{9}{47}x = 5.4 \rightarrow x = \frac{47 \times 5.4}{9} = 28.2$

**8) Answer: 9.**

$(2^a)^b = 512 \rightarrow 2^{ab} = 512$

$512 = 2^9 \rightarrow 2^{ab} = 2^9 \rightarrow ab = 9$

**9) Answer: A.**

If 14 balls are removed from the bag at random, there will be one ball in the bag.

The probability of choosing a white ball is 1 out of 15. Therefore, the probability of not choosing a white ball is 14 out of 15 and the probability of having not a white ball after removing 14 balls is the same.

**10) Answer: 7.**

The value of $y$ in the $x$-intercept of a line is zero. Then:

$y = 0 \rightarrow 5x - 8(0) = 35 \rightarrow 5x = 35 \rightarrow x = \frac{35}{5} = 7$

then, $x$-intercept of the line is 7

**11) Answer: B.**

Use distributive property:

$7x^3y(6x + 3y) = 7x^3y(6x) + 7x^3y(3y) = 42yx^4 + 21x^3y^2$

**12) Answer: 67.2.**

$160 \times \frac{42}{100} = 67.2$.

**13) Answer: D.**

$y = -3x + 2$

$(1, -1) \Rightarrow -1 = -3(1) + 2 \Rightarrow -1 = -1$

$(3, -7) \Rightarrow -7 = -3(3) + 2 \Rightarrow -7 = -7$

$(-2, 8) \Rightarrow 8 = -3(-2) + 2 \Rightarrow 8 = 8$

$(-3, 7) \Rightarrow 7 = -3(-3) + 2 \Rightarrow 11 \neq 7$

**14) Answer: 276.**

The perimeter of the trapezoid is 64 cm.

# ACT Aspire Subject Test Mathematics Grade 8

Therefore, the missing side (height) is = 64 − 17 − 14 − 9 = 24

Area of a trapezoid: A = $\frac{1}{2}$ h (b1 + b2) = $\frac{1}{2}$ (24) (9 + 14) = 276

**15) Answer: A.**

A graph represents $y$ as a function of $x$ if $x_1 = x_2 \rightarrow y_1 = y_2$

In choice A, for each $x$, we have two different values for $y$.

**16) Answer: B.**

$-25 < -7x - 4 < -4 \rightarrow$ Subtract 4 to all sides.

$-25 + 4 < -7x - 4 + 4 < -4 + 4$

$\rightarrow -21 < -7x < 0 \rightarrow$ Divide all sides by $-7$.

(Remember that when you divide all sides of an inequality by a negative number, the inequality sing will be swapped. < becomes >)

$\frac{-21}{-7} < \frac{-7x}{-7} < \frac{0}{-7} \Rightarrow 3 > x > 0$, or $0 < x < 3$

**17) Answer: B.**

Use simple interest formula: $I = prt$ (I = interest, p = principal, r = rate, t = time)

I = (24,000)(0.0195)(5) = 2,340

**18) Answer: A.**

Compare each mark:

In Algebra Joe scored 43 out of 33 in Algebra. It means Joe scored 76.74% of the total mark. $\frac{33}{43} = \frac{x}{100} \Rightarrow x = 76.74\%$

Joe scored 26 out of 51 in science. It means Joe scored 50.98% of the total mark.

$\frac{26}{51} = \frac{x}{100} \Rightarrow x = 50.98\%$

Joe scored 18 out of 29 in mathematic that it means 62.06% of total mark.

$\frac{18}{29} = \frac{x}{100} \Rightarrow x = 62.06\%$

Therefore, his score in Algebra is higher than his other scores.

**19) Answer: 384 $m^3$.**

Use the volume of the triangular prism formula.

V = $\frac{1}{2}$ (length) (base) (high) → V = $\frac{1}{2} \times 16 \times 8 \times 6 \Rightarrow$ V = 384 $m^3$

# ACT Aspire Subject Test Mathematics Grade 8

**20) Answer: D.**

To find the discount, multiply the number by (100% − rate of discount).

Therefore, for the first discount we get: $(100\% - 26\%)(E) = (0.74)E$

For increase of 7 %:

$(0.74)E \times (100\% + 7\%) = (0.74)(1.07) = 0.792E$.

**21) Answer: C.**

$3x - 5y = -14 \Rightarrow -5y = -3x - 14 \Rightarrow y = \frac{-3}{-5}x - \frac{14}{-5} \Rightarrow y = \frac{3}{5}x + \frac{14}{5}$

$y = \frac{3}{5}x + 2\frac{4}{5}$

**22) Answer: C.**

When a point is reflected over $x$ axes, the $(y)$ coordinate of that point changes to $(-y)$ while its $x$ coordinate remains the same. C $(4, 7) \rightarrow$ C' $(4, -7)$

**23) Answer: A.**

Solving Systems of Equations by Elimination

$\begin{array}{l} 7x - 10y = 9 \\ 2x - 3y = 3 \end{array}$ Multiply the first equation by 2, and second equation by −7, then add two equations.

$\begin{array}{l} 2(7x - 10y = 9) \\ -7(2x - 3y = 3) \end{array} \Rightarrow \begin{array}{l} 14x - 20y = 18 \\ -14x + 21y = -21 \end{array} \Rightarrow y = -3.$

$2x - 3y = 3, 2x - 3(-3) = 3$, then: $x = -3, (-3, -3)$

**24) Answer: B.**

To find the discount, multiply the number by (100% − rate of discount).

Therefore, for the first discount we get: $(410)(100\% - 23\%) = (410)(0.77)$

For the next 9 % discount: $(410)(0.77)(0.91)$.

**25) Answer: 13.**

Use distance formula:

$C = \sqrt{(x_A - x_B)^2 + (y - y_B)^2} \Rightarrow C = \sqrt{(7 - 2)^2 + ((-8) - 4)^2}$

$C = \sqrt{(5)^2 + (-12)^2} \Rightarrow C = \sqrt{25 + 144} \Rightarrow C = \sqrt{169} = 13$

# ACT Aspire Subject Test Mathematics Grade 8

**26) Answer: C.**

Percentage of men in city C = $\frac{760}{1,430} \times 100 = 53.15\%$

Percentage of women in city B = $\frac{290}{650} \times 100 = 44.62\%$

Percentage of men in city C to percentage of women in city B: $\frac{53.15}{44.62} = 1.19$

**27) Answer: D.**

Ratio of women to men in city A: $\frac{540}{710} = 0.76$

Ratio of women to men in city B: $\frac{290}{360} = 0.81$

Ratio of women to men in city C: $\frac{670}{760} = 0.88$

Ratio of women to men in city D: $\frac{530}{590} = 0.90$

**28) Answer: C.**

Solve for each equation: $(3, -7)$

$y = -7 \Rightarrow -7 = -7$

$y = -4x + 5 \Rightarrow -7 = -4(3) + 5 \Rightarrow -7 = -7$

$y = 8 - 3x \Rightarrow -7 = 8 - 3(3) \Rightarrow -1 \neq -7$

$y = -5x + 8 \Rightarrow -7 = -5(3) + 8 \Rightarrow -7 = -7$

**29) Answer: 18.**

$a + b + c = 46 \Rightarrow \frac{a+b+c+d}{4} = 16 \Rightarrow a + b + c + d = 64$

$\Rightarrow 46 + d = 64 \Rightarrow d = 64 - 46 = 18$

**30) Answer: B.**

$19 \times \$167 = \$3,173$

Payable amount is: $\$5,048 - \$3,173 = \$1,875$

**31) Answer: C.**

If two lines are parallel with each other, then the slope of the two lines is the same.

Then in line $y = 9x - 4$, the slope is equal to 9

And in the line $9x - y = -4 \Rightarrow y = 9x + 4$, the slope equal to 9

# ACT Aspire Subject Test Mathematics Grade 8

**32) Answer: D.**

Let $x$ be the smallest number. Then, these are the numbers:

$x, x+1, x+2, x+3, x+4, x+5$

average $= \frac{\text{sum of terms}}{\text{number of terms}} \Rightarrow 39 = \frac{x+(x+1)+(x+2)+(x+3)+(x+4)+(x+5)}{6}$

$\Rightarrow 39 = \frac{6x+15}{6} \Rightarrow 234 = 6x + 15 \Rightarrow 219 = 6x \Rightarrow x = 36.5$

**33) Answer: 900.**

Let $x$ be the original price.

If the price of a laptop is decreased by 40% to $540, then:

$60\% \ of \ x = 540 \Rightarrow 0.60x = 540 \Rightarrow x = 540 \div 0.60 = 900$

**34) Answer: 27.**

Let $x$ be the number. Write the equation and solve for $x$.

$\frac{3}{7} \times 42 = \frac{2}{3} \cdot x \Rightarrow \frac{3 \times 42}{7} = \frac{2x}{3}$, use cross multiplication to solve for $x$.

$9 \times 42 = 2x \times 7 \Rightarrow 378 = 14x \Rightarrow x = 27$

**35) Answer: D.**

The sum of the weight of all girls is: $13 \times 35 = 455$ kg

The sum of the weight of all boys is: $22 \times 45 = 990$ kg

The sum of the weight of all students is: $455 + 990 = 1,445$ kg

average $= \frac{\text{sum of terms}}{\text{number of terms}}$ ; average $= \frac{1,445}{35} = 41.28$

**36) Answer: 88.**

$\frac{a+b+c}{3} = 56 \Rightarrow a+b+c = 168$

$\frac{a+b+c+d}{4} = 64 \Rightarrow a+b+c+d = 256 \Rightarrow 168 + d = 256 \Rightarrow d = 256 - 168 = 88$

**37) Answer: C.**

$x - y = 3$ has a graph that is a straight line. All other options are not equations of straight lines.

**38) Answer: 36.**

The sum of supplement angles is 180. Let $x$ be that angle. Therefore,

$x + 4x = 180$; $5x = 180$, divide both sides by 5: $x = 36$

# ACT Aspire Subject Test Mathematics Grade 8

**39) Answer:** $16\pi$.

Let P be circumference of circle A, then; $2\pi r_A = 32\pi \to r_A = 16$

$r_A = 4r_B \to r_B = \frac{16}{4} = 4 \to$ Area of circle B is; $\pi r_B^2 = 16\pi$

**40) Answer: B.**

The probability of choosing a Hearts or Spades or Hearts is $\frac{39}{52} = \frac{3}{4}$

**41) Answer: A.**

If the score of Harper was 96, therefore the score of Emma is 48. Since the score of Zoe was one sixth of Emma, therefore, the score of Zoe is 8.

**42) Answer: 14.**

The median of a set of data is the value located in the middle of the data set. Combine the 2 sets provided, and organize them in ascending order:

$\{4, 6, 7, 9, 13, 15, 17, 19, 23, 26\}$

Since there are an even number of items in the resulting list, the median is the average of the two middle numbers.

Median= $(13 + 15) \div 2 = 14$

**43) Answer: C.**

The smallest possible value of $f(x)$ will occur when $x = 0$. Since $x^3$ is always positive, any positive or negative value of $x$ will make the value of $f(x)$ greater than 7.

Substitute 0 for $x$ and evaluate the expression: $f(0) = 8(0)^3 + 7 = 7$

**44) Answer: 48.**

$(16x^2 - 25y^2) = (4x - 5y)(4x + 5y)$; Then: $16x^2 - 25y^2 = 6 \times 8 = 48$

**45) Answer: C.**

Area of a circle equals: $A = \pi r^2$

The new diameter is 30% larger than the original then the new radius is also 30% larger than the original.

30% larger than $r$ is $1.3r$. Then, the area of larger circle is:

$$A = \pi r^2 = \pi(1.3r)^2 = \pi(1.69r^2) = 1.69\pi r^2$$

$1.69\pi r^2$ is 69% bigger than $\pi r^2$.

# ACT Aspire Subject Test Mathematics Grade 8

**46) Answer: 1.**

Subtract $\frac{1}{8b}$ and $\frac{1}{4b^2}$ from both sides of the equation. Then:

$\frac{1}{8b^2} + \frac{1}{8b} = \frac{1}{4b^2} \to \frac{1}{8b^2} - \frac{1}{4b^2} = -\frac{1}{8b}$

Multiply both numerator and denominator of the fraction $\frac{1}{4b^2}$ by 2. Then:

$\frac{1}{8b^2} - \frac{2}{8b^2} = -\frac{1}{8b}$ ; Simplify the first side of the equation: $-\frac{1}{8b^2} = -\frac{1}{8b}$

Use cross multiplication method: $8b = 8b^2 \to b = b^2 \to b = 1$

**47) Answer: D.**

Use FOIL (First, Out, In, Last)

$(3x + 5)(x + 7) = 3x^2 + 21x + 5x + 35 = 3x^2 + 26x + 35$

**48) Answer: $64\ cm^2$.**

First draw an isosceles triangle. Remember that two sides of the triangle are equal. Let put $a$ for the legs. Then: Use Pythagorean theorem to find the value of $a$.

$a^2 + b^2 = c^2 \to a^2 + a^2 = 16^2$

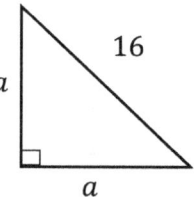

Isosceles right triangle

Simplify: $2a^2 = 256 \to a^2 = 128 \to a = \sqrt{128}$

$a = \sqrt{128} \Rightarrow$ area of the triangle is $= \frac{1}{2}(\sqrt{128} \times \sqrt{128}) = \frac{1}{2} \times 128 = 64\ cm^2$

**49) Answer: A.**

Factor each trinomial $x^2 - 5x - 24$ and $x^2 - x - 56$

$x^2 - 5x - 24 \Rightarrow (x - 8)(x + 3)$

$x^2 - x - 56 \Rightarrow (x - 8)(x + 7)$

The common factor of both expressions is $(x - 8)$.

**50) Answer: 7.**

Plug in the values of $x$ and $y$ in the equation:

$3 \blacksquare 5 = \sqrt{3^2 + 8(5)} = \sqrt{9 + 40} = \sqrt{49} = 7$

**51) Answer: A.**

First, convert all measurement to foot.

One foot is 12 inches. Then: 9 inches = $\frac{9}{12} = \frac{3}{4}$ feet

The volume flower box is length ×width ×height =$6 \times \frac{3}{4} \times 4 = 18$ cubic feet.

**52) Answer: D.**

$6x^5y^2 + 3x^4y^7 - (3x^5y^2 + 9x^4y^7) = 6x^5y^2 + 3x^4y^7 - 3x^5y^2 - 9x^4y^7 = 3x^5y^2 - 6x^4y^7$

**53) Answer: 34.5.**

$A = bh \Rightarrow A = 5 \times 6.9 = 34.5$

**"End"**

www.ingramcontent.com/pod-product-compliance
Lightning Source LLC
Chambersburg PA
CBHW080436110426
42743CB00016B/3184